A PRODUCTION GUIDE FOR YOUNG PEOPLE

A PRODUCTION GUIDE FOR YOUNG PEOPLE

by John LeBaron
& Philip Miller

illustrations by Mary Aufmuth

Prentice-Hall, Inc. Englewood Cliffs, N.J.

To Chris

Printed in the United States of America •J
Designed by Diana Hrisinko

Prentice-Hall International, Inc., London
Prentice-Hall of Australia, Pty. Ltd., Sydney
Prentice-Hall Canada, Inc., Toronto
Prentice-Hall of India Private Ltd., New Delhi
Prentice-Hall of Japan, Inc., Tokyo
Prentice-Hall of Southeast Asia Pte. Ltd., Singapore
Whitehall Books Limited, Wellington, New Zealand

10 9 8 7 6 5 4 3 2 1

Library of Congress Cataloging in Publication Data

LeBaron, John,
Portable video.

Includes index.
Summary: Combines practical information on production skills with more general discussions of the electronic principles behind video production processes. Includes activities and exercises in applying production techniques.
1. Video recordings—Production and direction—Juvenile literature. 2. Video tape recorders and recording—Juvenile literature. [1. Video tape recorders and recording. 2. Television—Production and direction. 3. Electronics] I. Miller, Philip
II. Title.
PN1992.95.L4 1982 384.55 82-7710
ISBN 0-13-686535-6 AACR2
ISBN 0-13-686519-4 (pbk.)

acknowledgments

Like most books, this book couldn't have been written without the help of many people. For this manuscript, our most important helpers have been the young people with whom we've worked over the years. They've given us the inspiration, the ideas, and the experience that are translated here into paragraphs and chapters. We have also benefited much from the good work of our colleagues: schoolteachers, media specialists, cable television program directors, college students, librarians, and parents. The children and adults who have guided our work are too numerous to recognize by name. If they are reading these pages, they know who they are. To them we say, "Thanks!"

Looking back at our work, there are some colleagues who, by their own example, have provided special assistance. Kit Laybourne, producer of the new public television series on the electronic media, *Media Probes*, has been a source of creative ideas for the many years we have known him. His work has been characterized by a passionate dedication to the welfare of young people and a determination that they understand and control the media forms which dominate their lives. Richard Earley is another such person. Working for many years with handicapped children, Dick has used video to help them learn, produce, and feel confident about themselves. It has been our privilege to know and work with these two outstanding individuals. Their influence is very much a part of this book.

In the preparation of this manuscript, Dexter Strong has been both a colleague and a friend. Dexter's contributions are most evident in Chapters 2 and 3 and in the Glossary. Dexter also handled several administrative details and provided valuable criticism from the moment we began work on this manuscript. Matthew LeBaron, John's eleven-year-old son, also read much of the manuscript as it was being typed—asking questions, making suggestions, and keeping us on the right track for the young readers we are trying to reach. Janet Pukanasis, who is more of a partner than a typist, worked countless evenings and weekends transforming atrociously unreadable handwriting into neatly typed pages.

Mary Aufmuth worked weekends and evenings transforming our rough sketches and diagrams into skillful finished drawings. At Prentice-Hall, editors Barbara Francis and Betsy Torjussen gave us encouragement and counsel, and designer Diana Hrisinko's creativity gives the book its graphic excellence. They kept us at our task in a supportive—not a demanding—manner. The speed and efficiency with which they handled our work was nothing short of miraculous.

Clint Clemens made his portfolio available to us for some of the book's illustrations. In seeking photos that could illustrate the best in visual composition, we were fortunate to find Clint, one of the best in his field. Rex McNamany and the staff at the Tech Video Center, a home video store in downtown Boston, freely put their time and their facilities at our disposal. Through them, we were able to keep up to date on new equipment developments and to correct errors in the manuscript before it was too late. Whatever errors that remain are ours, not theirs.

Those most deserving of our gratitude are the members of our families, the people who suffer most directly from the personal absences that result from too much attention to a manuscript. Thank you Faith, Matthew, and Jessie LeBaron for your tolerance and your participation in this project. And thank you Anne for your continuing trust and faith and for your patience and support as you watched weekends slip away. Finally, we'd like to offer a special note of acknowledgment and admiration to Pam, Patty, Judy, and Mr. and Mrs. Philip Miller for their hard work and determination on the most important project of all.

contents

chapter 1

VIDEO: HOW IT STARTED AND HOW IT WORKS

2

A Short History

It is hard for us today to imagine life without television. But throughout history, and for the first part of this century, people lived their entire lives without television. Television as we know it didn't begin until the late 1940s, when the three major networks began TV broadcasting on a regular basis. In television's earliest years, pictures were fuzzy, screens were tiny, and costs were enormous. In several of the first television shows, actors and actresses had to wear green makeup and purple lipstick so they could be seen on the tiny black-and-white screens!

In 1941 NBC created the first experimental television "network." Television signals were relayed from studios in New York City to stations in Schenectady, New York, and in Philadelphia. However, the picture quality of these long-distance transmissions was not very good, and they were soon discontinued.

Later in the forties, the big national radio networks—NBC, CBS, and ABC—recognized that TV would be much bigger than radio, and they hurried to start their own TV broadcast services. NBC was the first, entering TV broadcasting mainly to sell television sets made by its parent company, RCA. NBC's earliest broadcasts were transmitted "live" and were received only in New York City (Fig. 1.1).

In the late 1940s and early 1950s, commercial TV network broadcasting began in earnest. The three major networks began to sign up local TV stations across the country as network affiliates. Programs originated at network headquarters in New York and were then sent to the affiliated stations over special telephone cables. The network system worked well, except for the stations on the West Coast, where the programs from New York appeared at the wrong time. Because of the time-zone difference, programs scheduled for a 9:00 P.M. adult audience in New York would appear in San Francisco and Los Angeles at 6:00 P.M., when most families were just sitting down to dinner.

To cope with this problem, the West Coast stations first tried filming the New York broadcasts from a television screen, using a special movie camera called a *kinescope*. The stations would then develop the kinescope film and quickly prepare it for broadcasting three hours later. But there were major problems with kinescoping. The developing process was time-consuming and unreliable; the film could be used only once (it could not be erased and reused to record other programs); and, most importantly, kinescoping did not produce very good pictures.

THE NATIONAL ARCHIVES

Figure 1.1 *Family TV viewing: The way it used to be.*

The Birth of Videotape Recording What the networks needed was a way to record and play back programs cheaply and quickly. The answer was the videotape recorder (VTR). Unlike the kinescope, a videotape recorder uses an electronic process to record TV programs. Instead of exposing a chemically treated film to light, a VTR exposes a magnetic videotape to electronic signals. This ribbonlike tape is coated with tiny metallic particles. Inside the VTR, special electromagnets called *video recording heads* rearrange these metallic particles into a magnetic pattern on the moving videotape. The magnetic pattern is invisible and corresponds exactly to the electronic signals sent to the VTR by a video camera and a microphone.

VTRs can record images (video signals) and sound (audio signals) together onto the same tape. When the tape is played back, the same video heads that were used to record the audio and video signals as magnetic patterns on the tape change those patterns back into electronic signals again. These electronic signals are then sent to a TV set, which in turn changes them back into images you can see and sounds you can hear. Also, since videotapes simply hold a magnetic charge and do not undergo any chemical change, they can be ''erased'' and used over and over again—a great savings for the TV stations and networks.

One of the first VTRs used for broadcasting was developed by the British Broadcasting Company (BBC) in 1955. It was a huge machine that used tape reels that measured five feet across for a short half-hour program. The BBC videotape recorder needed these large reels because it used a video head that didn't move. With the video head standing still, the videotape had to move across the head at breakneck speed—200 inches per second (IPS).

In 1956 the Ampex Corporation (an American company) created a major breakthrough by developing a machine that operated at a much slower tape speed—only 15 IPS. In the new Ampex machine, both the heads *and* the tape moved. When the speed of the moving heads and the speed of the moving tape were added together, the relative tape-to-head speed (or writing speed) of the Ampex VTR was very close to that of the earlier BBC system, even though the tape in the Ampex machine moved at a much slower speed. With the slower tape speed, the Ampex VTR could use smaller-size tape reels. The tape on the reels was two inches wide, and the Ampex machine was called a *quadruplex* (or quad) VTR. In 1959 the first color quad VTR was introduced, and many color quad machines are still being used today by professional TV broadcasters.

Small-Format VTRs The next generation of VTRs cut down on the size and expense of the equipment by narrowing the width of the videotape. Compared to the quadruplex VTR, which used a two-inch-wide videotape, these newer VTRs used tape that was only one inch wide. For this reason they were called *small-format* machines. The term "small-format VTR" refers to any VTR that uses tape one inch wide or narrower. (Because the quad system uses two-inch-wide videotape, it is called a large-format VTR.)

The first small-format VTRs appeared in 1960, and they introduced a new way of recording strips of information on the videotape. Like the quad system, the small-format VTRs used video heads that moved. But unlike the video heads on the quad machines that moved straight across the tape, the video heads on the small-format machines moved across the tape on a slant and in the opposite direction to the movement of the tape (Fig. 1.2). This way, small-format machines could pack a relatively large amount of video information on a relatively narrow tape. This new design was called the *helical-scan* or *slant-track* video recording system, and it is used by almost all VTRs found in homes, schools, and businesses today.

Portable Video In 1967 the Sony Corporation of Japan introduced the first half-inch-format Portapak VTR. The Portapak was a black-and-white helical-scan machine that used half-inch-wide tape. It was the first portable battery-operated small-format VTR. It was also the first video recording system priced low enough for schools and other small organizations to buy. The early Portapak systems cost less than $2000 each and included a VTR, or "deck," as it is often called, a camera, and a battery charger.

Convenient Video: The Videocassette Recorder In 1972 Sony again broke new ground by marketing the first videocassette recorder (VCR). Before 1972, all videotape recorders were "open-reel" VTRs that had to be threaded by hand. This manual tape-threading was time-consuming, and repeated handling of the tape made it easy to damage both the videotape and the video heads. For many years, though, these open-reel machines were the only low-cost portable VTRs available. In spite of their drawbacks, open-reel VTRs were reliable and durable, and many remain in use today, especially in schools.

Figure 1.2 ***Two major video recording systems.***

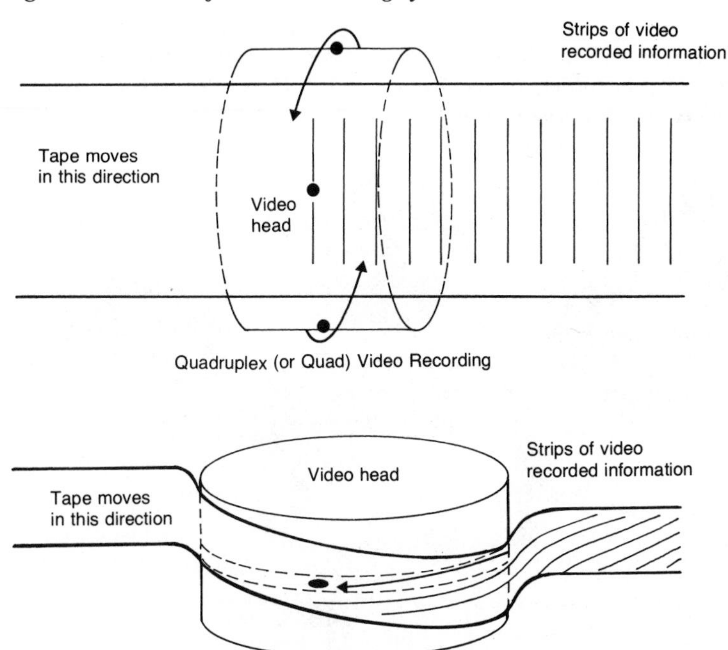

Sony's new VCR used the same basic helical-scan process as open-reel VTRs, but both the videotape feed and take-up reels on this new system were contained inside a rectangular plastic casing, called a cassette (Fig. 1.3). To load the videotape, you simply turned on the VCR's power and inserted the cassette. The tape and heads never had to be touched, and both were well protected from dust and dirt. The first VCRs used videotape that was ¾ inch wide, and they threaded the tape in a U-shaped pattern inside the deck. Because of this, they were called "U-type" or "¾-inch" VCRs.

Video for the Home By the mid-1970s, schools and businesses had found so many uses for video equipment that equipment manufacturers began to believe that video recorders might also be popular in homes. Again it was Sony who made the first move by introducing its half-inch Betamax (or Beta format) videocassette recorder for the home market in 1975. The Beta VCR uses a helical-scan head mechanism and videocassettes with half-inch-wide tape. When it first appeared, the Beta system was cheaper, smaller, and easier to operate than anything else available. Quick to recognize a good thing, the Japan Victor Corporation (JVC) introduced its own type of half-inch videocassette

Figure 1.3 ***Comparison of tape sizes: in the center, ¾" videocassette. Clockwise from upper left-hand corner: ½" VHS videocassette, ½" open-reel videotape, ½" Beta videocassette, ¼" Technicolor videocassette, standard audio cassette.***

Table 1.1

Major Brands of Beta Format Video Recorders	Major Brands of VHS Format Video Recorders
Sanyo	Akai
Sears	Hitachi
Sony	JVC
Toshiba	Magnavox
Zenith	Panasonic
	Quasar
	RCA
	Sharp

system, the VHS (Video Home System), in 1976. Soon a number of other manufacturers (such as Panasonic, RCA, and Zenith) also began making and selling their own versions of Beta or VHS systems.

Today the Beta and VHS formats pretty well dominate the home VCR market. Other home formats have come and gone, but most manufacturers who now make home machines make them in either the Beta or the VHS standard (Table 1.1). Although the Beta and VHS systems work in much the same way, they are not *compatible* with each other. A tape made on a VHS machine cannot be played on a Beta deck, or vice versa.

The first *portable* Beta and VHS decks appeared in 1978. Since then, these battery-powered half-inch VCRs have become smaller and even easier to use. The newest Beta and VHS portable decks can produce a picture quality nearly equal to three-quarter-inch portable decks, but they are much lighter and easier to carry. Since the birth of the home video market in 1975, color video cameras have also become much cheaper, much smaller, and much better. Today it's possible to buy a good color camera for less than $1000, and the price is still going down. In 1975 a color camera of the same quality would have cost more than $5000.

Looking Ahead By the time you read this book, there may be newer and smaller VCR formats on sale at video equipment stores. In fact, a company called Technicolor already sells a VCR that uses a quarter-inch cassette roughly the same size as an ordinary audio cassette (Fig. 1.3). Not to be outdone, other manufacturers are predicting a new type of VCR that will use cassettes the size of a matchbox! Also, as video-recording equipment gets smaller, manufacturers will begin combining the deck and the camera in the same housing. When these combined video systems become available, you won't need to carry a separate camera and deck. They'll both be part of the same small package.

You may have heard or read news reports about a "home video revolution." In fact, what we're experiencing today is more than just a home *video* revolution. We're actually in the middle of a much bigger "home electronic information revolution," a revolution brought on by changes in the number and types of TV services available to us. In addition to video-recording equipment that allows us to create our own TV programs, we can now buy videodisc players that allow us to play back video "records," and microcomputers that allow us to play video games and write our own computer programs. At the same time, cable television systems in some cities are delivering 100 or more channels to home subscribers, and several companies are planning to build satellite systems that will broadcast TV programs directly to homes through earth stations* costing less than $500. In the not-too-distant future, new types of printed (or alphanumeric) information will also be available for display on our home TV sets through telephone wires and cable TV lines or from ordinary TV broadcast stations.

The home electronic information revolution is here, and it will surely change the way we work, learn, and spend our leisure time. By learning about the new electronic equipment and making smart choices, we can help make sure that the changes in our own lives are changes for the better.

The Video Recording Process

People of all ages spend a lot of time in front of their television sets, but most give very little thought to what makes it possible for a television program to be sent from a TV station to the TV receiver in their homes. The same is true for videotape recording. Many people use VTRs, but very few know how they really work.

In some ways, videotape recording is like photography or film-making. In other ways, it is like recording sound on audio tape. In this section, you will learn how a video camera changes a picture into an

*An *earth station* is an antenna system capable of receiving TV signals directly from a communications satellite located 22,300 miles above the earth's equator.

electronic signal, how the videotape recorder stores and plays back that signal, and how a TV set changes the signal back into a picture.

Parts of a Portable Video System You need two basic pieces of video equipment to record images and sounds: a camera with a built-in microphone and a deck. To provide power to the camera and deck, you also need a battery or an AC adapter/battery charger. To play back a recorded videotape, you'll have to add a TV set. These components form an electronic pathway for the audio and video signals that create a video recording. The pathway begins at the camera, where light and sound are changed into electronic audio and video signals. The camera sends these audio and video signals to the video deck, which records and stores the signals for later playback. When you play back the recording, the deck sends the combined audio and video signals to a TV set—the final stop on the electronic video pathway. On the newest video systems, the camera and the deck may be housed in the same casing. Nonetheless, they are electronically separate, just as they would be if they were in separate housings.

The Portable Video Camera A video camera has two basic parts: a *lens* and a *body* (Fig. 1.4). The base of the lens is attached to the front of the camera body. The lens itself consists of several glass elements contained within a *lens barrel*. When we speak of a lens, we usually mean the lens barrel and the glass elements together.

The lens is the "eye" of the camera and works a lot like our own eyes. In a video camera, the lens focuses all the light it "sees" into the camera's body. Inside the camera's body there is a special pickup tube, often called a *vidicon tube*. The front surface of the pickup tube is flat and *photosensitive*, which means that it reacts to the varying brightnesses of light within the area the camera lens "sees." This flat photosensitive surface is called the target of the tube. In the human eye, the photosensitive target surface is called the retina. In both the video camera and the human eye, the light striking the target is changed into electrical currents. These electrical currents correspond directly to the image the eye or camera is seeing. Bright spots in the image appear as high points, or "peaks," in the electrical current, while dark spots appear as low spots, or "valleys." In humans, these electrical currents are sent from the eye to the brain. In a video system, the video camera sends the signals to a VTR deck for recording.

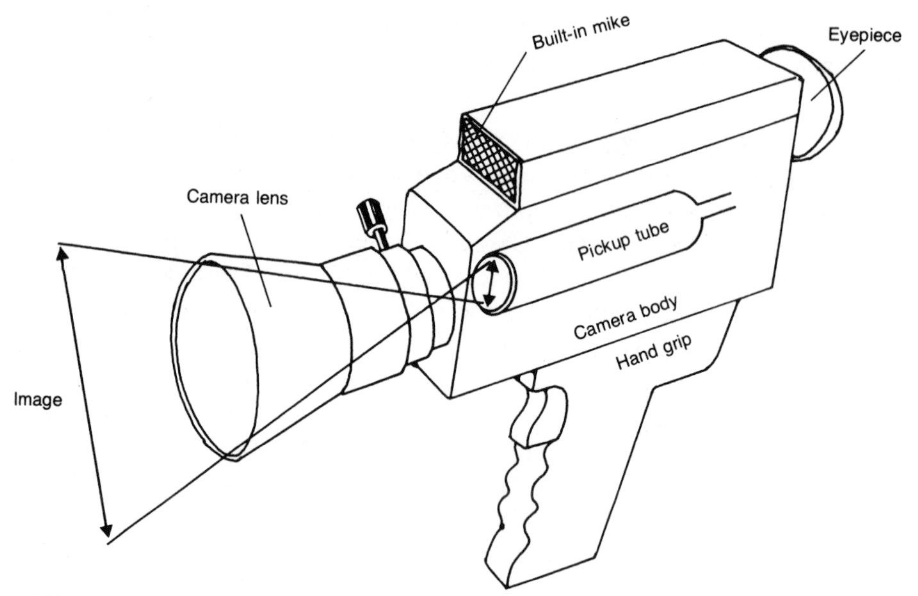

Figure 1.4 ***Parts of a video camera.***

Some of the newest color video cameras don't use a vidicon-type pickup tube. Instead of a pickup tube, these cameras use a small semi-conductor chip to sense an image's varying degrees of brightness as it passes through the camera lens. Cameras that use these semi-conductor chips are called *charge-coupled device* (CCD) or *metal-oxide semi-conductor* (MOS) cameras, and they offer several important advantages over tube-type cameras. They tend to be smaller and lighter than their tube-type counterparts, they eliminate the danger of "vidicon burn," and they do not "streak" after being pointed at a bright light source. They also boast excellent image quality. Eventually, chip cameras will be easier and cheaper to build than conventional tube-type cameras. Right now, however, low-priced CCD/MOS cameras haven't been available long enough to be able to judge their performance. But if they're as good as the manufacturers say they are, the days of the pickup tube camera are probably numbered.

The Video Camera Scanning System Inside a video camera's vidicon tube, an amazing process takes place. The pickup tube changes the varying brightnesses of light viewed by the camera lens into varying electrical currents. However, to produce a viewable image, these currents have to be read, or scanned, in an organized way. This scanning is

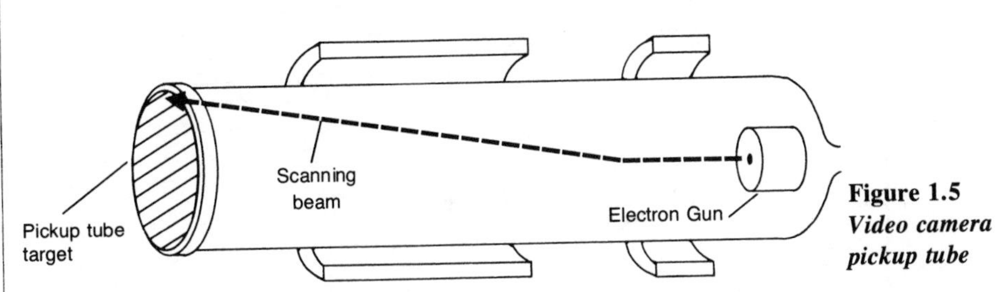

Figure 1.5
Video camera pickup tube

done by an *electronic gun* located at the far end of the pickup tube, opposite the target area (Fig. 1.5). The electron gun shoots a steady scanning beam of tiny electrons at the inside surface of the pickup tube's target, sweeping across it from left to right in a series of lines. The scanning beam moves from the top of the target to the bottom.

The camera lens focuses a picture on the target of the pickup tube and the photosensitive target changes the light and dark areas of that picture into varying degrees of electrical current. The scanning beam "reads" the current produced by those light and dark areas and changes this light and dark image information into organized electronic signals. This organized electronic current is the *video signal* sent from the camera to a VTR or TV set.

Like movies, video and television images are made from a series of still picture frames. In North American TV systems, each frame contains 525 lines of video information. Therefore, every time the scanning beam in a pickup tube scans 525 lines of information from top to bottom, it has scanned one full frame of video. The scanning beam then returns to the top left-hand corner of the pickup tube's target and begins scanning the next frame. There are thirty of these frames in every second of video recording or playback time.

But there is one odd wrinkle to this scanning process. A scanning beam does not scan every line of video information as it moves from the top to the bottom of the image. What it does first is to scan only the odd-numbered lines (262.5 of them) from top to bottom to create one *field* of video information. Then the beam returns to the top of the target again and scans the even-numbered lines. These even-numbered lines make up the second field. It takes two fields to make one frame. Since a frame is scanned in 1/30 of a second, it takes half that time (1/60 of a second) to scan a field (Fig. 1.6). This system sounds strange, but there is a reason for it. The scanning beam becomes weaker as it moves from the top to the bottom of the pickup tube's target. If all 525 lines were scanned in order, the strength of the beam would be so weak at the bottom it would not read the image accurately. When it was played back,

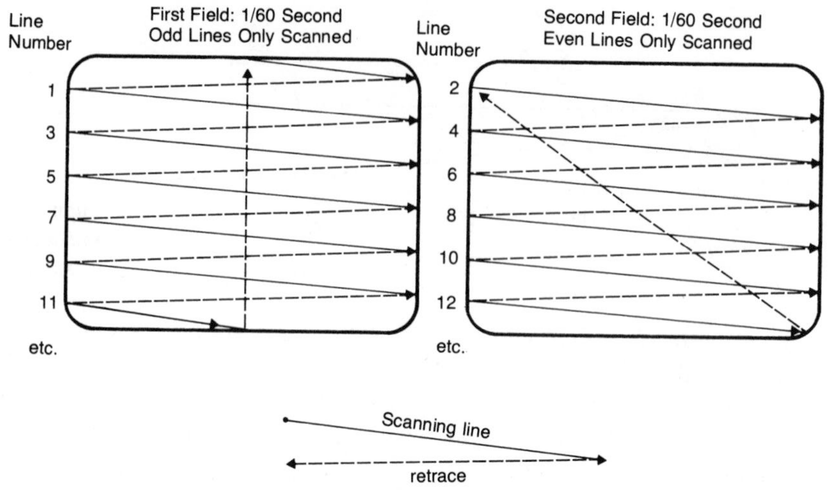

Figure 1.6 ***The television scanning process: two fields (1/60 second each) = one frame (1/30 second).***

this weak signal would cause the image on the television screen to flicker.

How does the scanning beam know when to begin scanning each new line and when to return to the top of the target to begin a new field of information? A separate electronic signal, called a *sync* (or synchronizing) *signal,* controls the beam and keeps it in line. The sync signal is divided into two parts: horizontal and vertical sync. The horizontal-sync pulses control the line-by-line scanning process. The vertical-sync pulses tell the electron gun when to begin each new field of video. Without sync signals, a video signal is useless. It would display chaotic and unviewable video information on the television screen. A usable signal that combines video and sync signals is called a *composite video signal.*

The Television Set Most of the television pictures you see were originally created by a video camera.* Some of these images may come to your TV set from a VTR that is sending the signal from a television station, but a video camera probably created the signal in the first place.

The television set works like a video camera in reverse. The TV picture tube, called a *cathode ray tube,* or CRT, is much like the video camera's pickup tube. Both have electron guns and both shoot a stream of electrons at a photosensitive surface. The inside front surface of the TV picture tube is called the *raster* area (Fig. 1.7). The outside surface of the TV picture tube is the TV screen.

*Television pictures can also be created by computers and other electronic machines.

The main difference between the camera and TV tube is that the scanning beam in the camera's pickup tube "reads" information coming in from a camera lens, while the scanning beam in a TV picture tube "writes" information onto the raster area. The raster area is coated with countless numbers of tiny *phosphor dots*. These phosphor dots glow brightly or dimly, according to the strength of the electron beam striking it at any particular moment. The stronger the beam, the brighter the dots. The strength of the electron beam in the TV set depends on the strength of the video camera's electron beam when it originally "read" the video image. If the camera's electron beam is strong at a certain spot, the television set's electron beam will be strong at that point, too. The sync signals sent from the camera are important here, too. The sync signals coming from the camera make sure that the electron gun in the TV's picture tube scans the raster area in exactly the same way that the pickup tube's electron beam scanned the camera's target area. If everything is working as it should, the original picture viewed by the camera should appear as a clear, viewable image on your TV screen.

The Videotape Recorder In a video recording system, the videotape recorder (VTR) comes between the camera and the television set. The VTR stores the image scanned by the pickup tube, allowing you to watch it at a later, more convenient time. The VTR is the device that records and stores the electronic code that the video camera is "seeing." The VTR can then play back this recorded code for viewing and listening on a television set.

A VTR records images (video) and sounds (audio) at the same time. It also records a sync signal. As mentioned earlier, the sync signal determines the exact scanning pattern of the electron gun in the TV set's

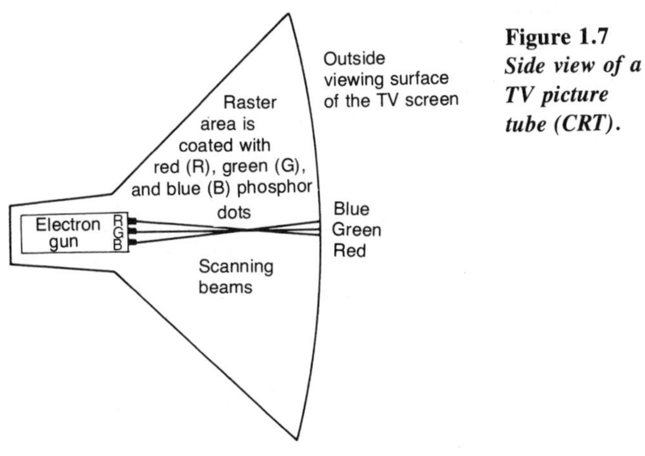

Figure 1.7
Side view of a TV picture tube (CRT).

Figure 1.8 *Magnified view of a video recording/playback head.*

picture tube when the videotape is played back. The video signal is recorded in slanted tracks as the videotape moves by the two video heads (or four heads in the newer cassette decks). The video heads protrude slightly through a slot that circles the side wall of a video head drum (Fig. 1.8), and the heads spin against the opposite movement of the tape. As the video heads sweep across the videotape surface, the heads place slanted strips of video information in the middle of the tape's width.

The outer edges of the tape are used for recording the sync signals and the audio signals (Fig. 1.9). Every VTR has sync and audio heads that don't move. They are usually located outside the video head drum (Fig. 1.10). A VTR also has one more head located at the beginning of the tape path, so the videotape passes by this head before it reaches the audio and video recording heads. This is the erase head. Its job is to erase all old information on a tape before any new information is recorded onto it. Without an erase head, you would not be able to erase and reuse videotapes.

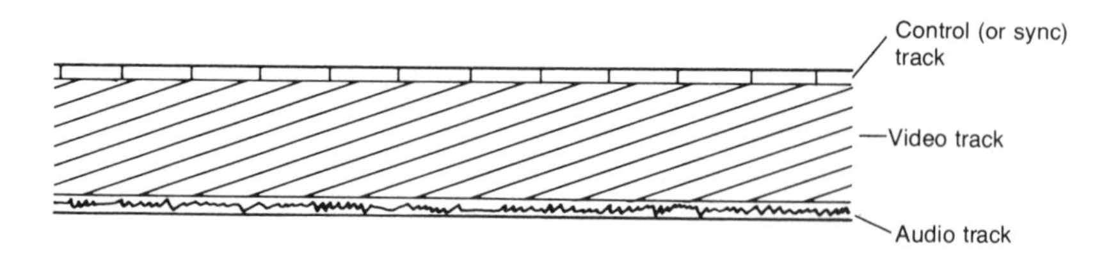

Figure 1.9 *Three information tracks on a videotape.*

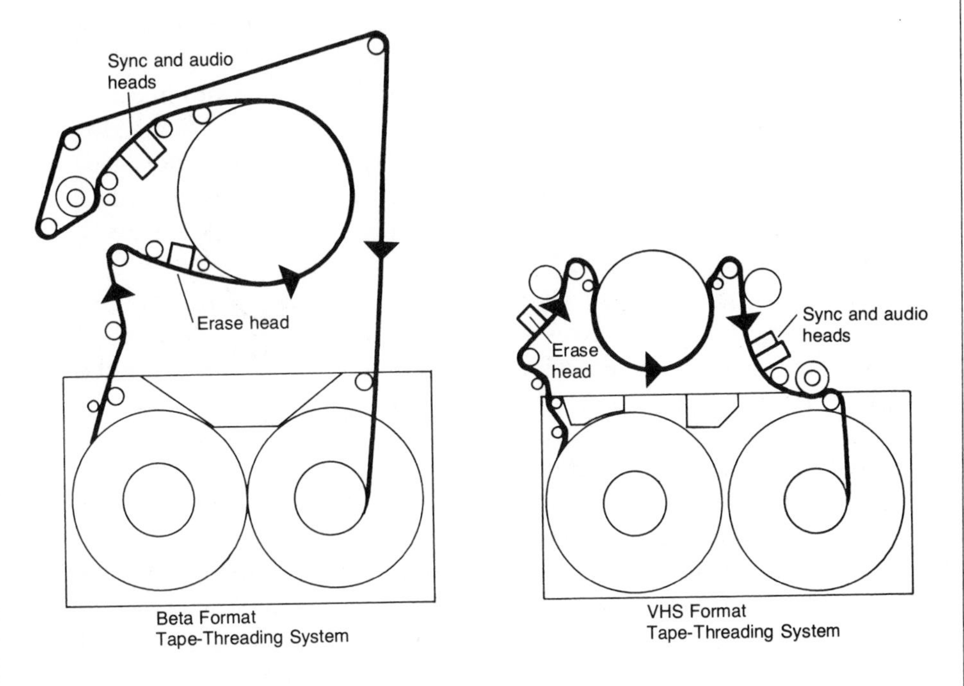

Figure 1.10 ***Beta and VHS videocassette tape-threading paths (not drawn to scale).***

The Videotape A videotape is a long ribbonlike strip composed of three layers. The first layer is a special kind of plastic base, often called mylar. This is the backing of the tape, used to hold the rest of the tape together. On the top of this plastic base is a gluelike surface called adhesive binder. This binder holds a coating of metal oxide (usually *ferrous oxide*) particles. These metallic particles are then ''magnetized'' into strips of information when swept by the spinning video recording heads of a VTR. The ferrous oxide particles are microscopic and cannot be seen by the naked eye. The recording side of the tape is shiny, while the reverse side is usually a dull black or brown.

In film-making, the difference between color and black-and-white production is in the film itself. All movie cameras are capable of making either color or black-and-white films; but to make color movies, the camera must be loaded with color film and to make black-and-white movies, you need black-and-white film.

In video, the electronic components determine whether a program will be in color or black and white, not the videotape. All videotape can record both in color and in black and white. As far as we know, all of the newer portable home videocassette recorders can also record in color. But in color video production, you also need a special color camera for recording and a color television set for playing back the color tape. These color video components use the same scanning process described earlier in this chapter, but they include extra parts and circuits that allow them to record and play back color video signals.

If you take good care of your videotapes, you can erase and re-use them many times. This gives you much more opportunity to practice your shots and experiment with program production. If you don't like what you shot the first time, you can try again on the same tape. But film can only be exposed once. Once it's shot, it's shot forever. Some videotape manufacturers claim that a single tape can withstand up to 500 trips or "passes" through a VTR. (A pass includes recording, playing back, rewinding, and fast forwarding.)

If you're still confused about the way video recording and playback work, take a look at the game that is described in the next section of this chapter. Get two or three friends or family members together and try it out.

A Game to Simulate the Video Recording Process

The Human Video System Understanding how video and television work is hard for some people. Here's a simple game that's been designed to help you understand the video scanning and transmission process. In the game, you'll work with a model of the scanning process. The model does not show exactly what happens inside a video camera and a TV set, but it's close enough to make the point. You can play this game with readily available supplies, and you can easily make all the game pieces yourself. After you've made the pieces, the game will take about twenty minutes to play.

Earlier in this chapter, you learned that the video system works on the basis of a special electronic code. This code is the picture translated into an electronic signal, a signal which can then be sent over electronic cables to a VTR or to a TV set. The most basic kind of video system doesn't need a videotape recorder at all. It consists only of a video camera connected directly to a television set. The television set shows an image at the same time that the video camera "sees" it, so this basic video system can only show "live" pictures. This type of camera/TV system is often used in security systems in stores or banks.

In this game, you too will be working in a special code. Your code will be a translation of a real image, just like the electronic video signal sent from a video camera. By using this code, you and a friend will be able to reconstruct a "video image" on a "TV screen" miles away from the "camera." When you play this game, you'll be creating a simple model of a real video transmission. This kind of model is called a *simulation,* and the game-players will simulate the actual functions of the scanning process in a video camera and a television set. The game may be played by two or three people.

Here are the materials you'll need to construct this game:

- a sheet of clear transparent plastic (or *acetate*)
- several sheets of plain white 8½″ x 11″ paper
- one black felt-tipped marker with a fine point capable of marking on clear plastic (a pen used to write on the overhead projector transparencies used in schools will do)
- three broad-tipped markers: red, blue, and green (these *don't* have to be the kind that write on plastic)
- a ruler
- some paper clips
- two telephones for an ordinary local call (optional)
- some self-adhesive name badges (optional)

First, you'll need to build the pieces of the game. On one of the blank pieces of paper, draw a simple rectangle—3″ high by 4″ wide. Underneath this rectangle, write the words "REAL-LIFE IMAGE" as shown in Figure 1.11a. This will become the "image" that an imaginary camera lens will focus onto the simulated vidicon tube target.

The camera pickup tube target is drawn on the piece of clear acetate, which can be a standard-sized overhead projector transparency or any other piece of clear plastic cut to a size no smaller than 8½″ x 11″. On this transparent acetate, you should draw another 3″ x 4″ rectangle with a fine-tipped black marker so that it lines up precisely with the 3″ x 4″ real-life image rectangle you just drew on the paper.

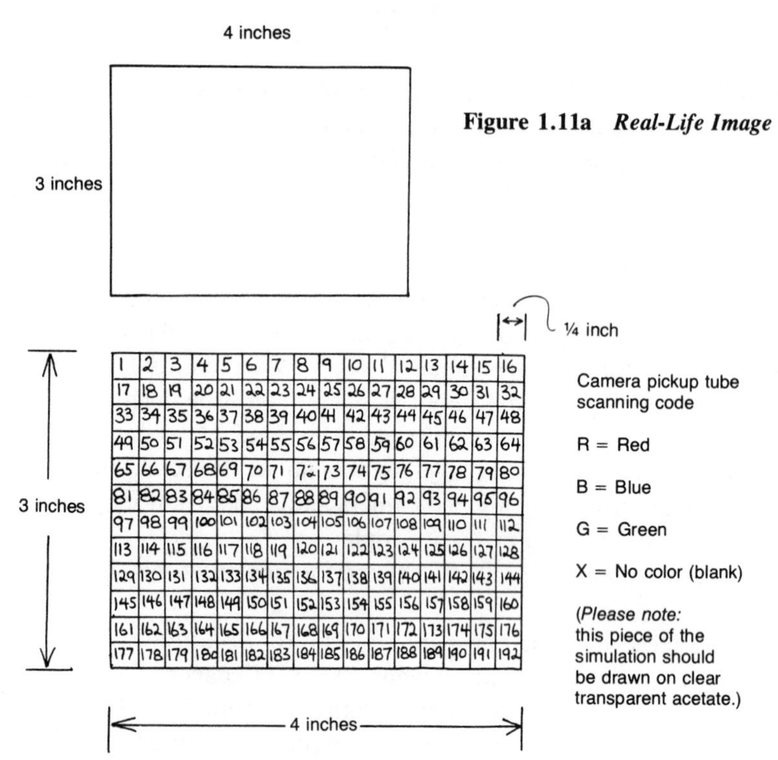

Figure 1.11a ***Real-Life Image***

Figure 1.11b ***Target of Video Camera's Pickup Tube (Inside Surface)***

The rectangle you just drew on the acetate will outline the pickup tube target area, and you'll need to divide it into small quarter-inch squares, 16 across by 12 down. You should number each small square in order from 1 to 192, starting at the upper left-hand corner and proceeding line by line to the lower right-hand corner. This is shown in Figure 1.11b. Under the 3″ x 4″ rectangle drawn onto the acetate, write the words "VIDEO CAMERA'S PICKUP TUBE TARGET (INSIDE SURFACE)." To the right of this rectangle, write down the pickup tube's scanning code as shown in the illustration. (In a color video system, the scanning code uses the colors red, blue, and green. These are the three primary colors from which all other colors are made. A real color video camera only senses these three colors, mixing and matching them to produce all the colors you see on a TV screen.)

On the second piece of 8½″ x 11″ plain paper, draw the simulated TV picture tube raster area. This is also a 3″ x 4″ rectangle. Like the

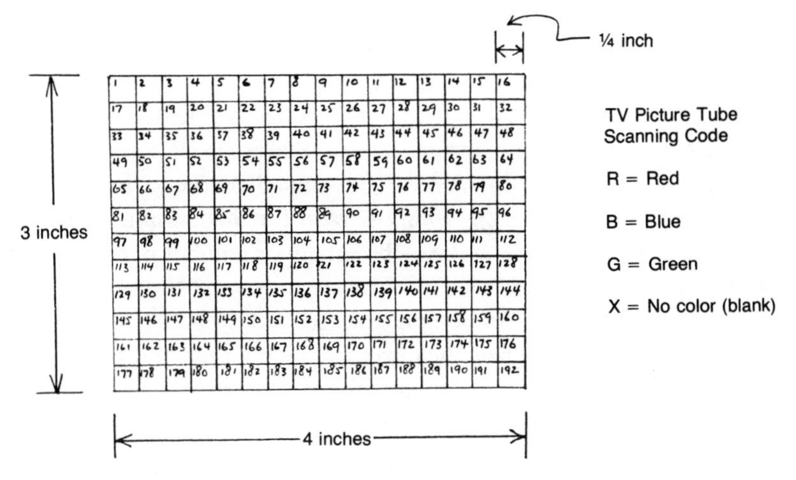

Figure 1.11c *TV Picture Tube Raster Area*

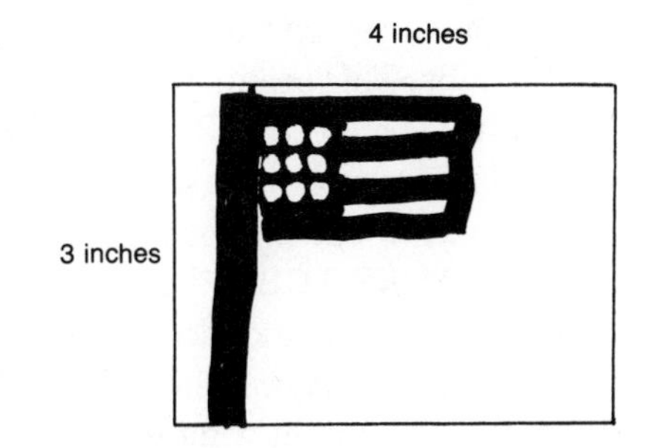

Figure 1.11d *Real-Life Image*

pickup target area you drew on the piece of clear acetate, it is divided into quarter-inch squares and numbered in order from 1 to 192 as pictured in Figure 1.11c. Write the words "TV PICTURE TUBE RASTER AREA" under the raster area rectangle and write the TV picture tube scanning code as shown.

These pieces take a fairly long time to draw. For that reason, if you or your friends are likely to do this simulation several times, it would be a good idea to make Xerox or ditto copies of the original drawings to use as working papers. Save the original ditto masters or drawings to make more duplicates when you need them.

Playing the Game You're now ready to start the game. Using the red, blue, and green broad-tipped markers, draw a very simple image inside the plain rectangle marked "REAL-LIFE IMAGE." The game works best if you draw the picture with thick vertical and horizontal lines. Don't use too many circles or diagonal lines. When drawing the image, use all three colors of the felt-tipped markers. Figure 1.11d shows a picture drawn by a young person for this simulation.

Either the person who drew the image or another person takes on the role of the video camera. (This person may want to mark the words "VIDEO CAMERA" on a name badge and wear it, but this isn't absolutely necessary.) Place the clear acetate over the real-life image so that the edges of the 3″ x 4″ pickup tube target line up *exactly* with the edges of the 3″ x 4″ real-life image. You can paper-clip the paper sheet and the clear acetate together so that they don't slip apart.

What you now have is a simulated real-life image focused on the target area of a camera pickup tube. The image should show through the quarter-inch squares of the pickup target.

You now need another person to simulate the role of the television picture tube. This person will be working with the piece of paper marked "TV PICTURE TUBE RASTER AREA" (the paper with the rectangle divided into quarter-inch squares). To avoid any possible confusion, this person may wish to wear a name badge. The name badge should say "TV PICTURE TUBE." To carry out the job properly, work with red, blue, and green markers.

If you're playing the role of the video camera with a friend or neighbor, you might want to send (or transmit) the image you scan by telephone. To do this, the person simulating the video camera calls the person performing as the television picture tube (so the telephone circuit between your house and your friend's house would simulate the cable connecting a video camera to a television set). A telephone isn't absolutely necessary, though. You can play the game just as well with two people in different parts of the same room.

Let's say that you are simulating the video camera. As we said, you begin by placing the acetate "pickup tube target" over the paper "real-life image" so they line up together. Then your eye becomes the simulated electron gun of the pickup tube, and your line of sight becomes the scanning beam. You scan each of the twelve lines of image information striking the pickup tube target, square by square. (Remember, though, that if you were a real scanning beam, you'd scan the odd-numbered lines first. Then, you'd return to the top of the target area to scan the even-numbered lines.)

As you scan, you transmit the picture information by code. If square 1 is blank, you call out "one, X." If square 2 is red, call "two,

R.'' If square 3 is blue, call ''three, B,'' and so on until you reach the end at square 192. If a square is partially occupied by one color and partly by another, call the main color. *Do not* say something like ''four, partly B and partly R.'' If two colors appear equally in a single square, use your best judgment, but call *only one* color.

As you do this, your partner will be filling the squares on the TV picture tube raster area in the same code as the one you're using. Your partner will perform this job square by square, number by number, exactly as you direct. In effect, you're sending your partner a composite video signal, because you're sending him information about what's in the image and information about the way the image is being scanned. (Remember, a composite video signal contains both video and sync information.)

When your partner plays the role of a TV picture tube, his hand acts as the picture tube's electron gun and the felt-tipped markers simulate the scanning beam that causes the different-colored phosphor dots on the raster area (which are represented by the quarter-inch squares) to ''glow'' as you've directed.

At the end of the game, you can compare the image produced by your partner with the original real-life image. Your partner's TV picture tube raster will display a three-color image that will be a very rough duplicate of the real-life image that was ''read,'' scanned, and transmitted by the video camera. It will be rough, because in this activity a small number of very broad lines are scanned, producing a very crude copy of the original image. Remember, the real video-scanning process scans 525 lines, not the twelve lines you scanned in this simulation. Not only that, but when you play this game, you translate twice—once from a picture into a transmission code, and then from the code back into a picture again, just as a real video transmission system does. So don't be disappointed if your simulated TV picture doesn't look exactly like the real-life image. It's not supposed to! If you had the two or three years it would take to scan 525 lines by hand with paper and felt-tipped markers, your simulated TV picture would look as sharp and accurate as the best color TV pictures.

Now play the game and see if the scanning process becomes any clearer to you.

chapter 2

GETTING READY: COLLECTING AND CONNECTING THE PIECES OF A PORTABLE VIDEO SYSTEM

Getting Your Equipment Together

To begin producing video programming, you'll first need to locate some video equipment. Perhaps your family, or friends of your family, have bought a portable video system for home use. Or maybe a neighborhood school has video equipment you could borrow. If the town that you live in has cable television, the local cable company may have a portable video system that you can use. In many towns and cities, the cable company is required by its contract to provide this equipment for people like you. Call your local cable company and ask the program director how to go about borrowing the cable system's portable video equipment.

Whether you own or borrow video recording equipment, you should make sure you have all the parts you'll need to record and play back programming. There are four basic parts (or components) to any portable video system:

1. the video camera
2. the videotape recorder (VTR), sometimes called a deck
3. the power supply
4. a television set

To connect the components, you will also need the proper *connecting cables*.

Like most newcomers to video, you will probably need help setting up, connecting, and operating the components of your portable video system. The operator's manual that comes with most video systems is a good place to start. The manual can tell you a lot about the way your video system works, and it's always good to have on hand. But the operator's manual may not give you all the information you need, and it may not always make complete sense to you. That's where this chapter will help. It will explain the pieces of a video system and tell you how to connect the pieces together for video recording.

The Electronic Pathway

When your portable system is properly connected, the video camera, videotape recorder (VTR), and TV set form an electronic pathway for the signals you need to create a videotape recording. The electronic pathway starts with the camera, passes through the VTR, and ends at the TV set. A fourth component—the power supply—provides electricity to power the VTR and the camera. (See Figure 2.1.)

In Chapter 1, you learned how each component of the video system performs its special task along the videotape recording pathway. Now we'll describe *what* each component does and what *you* must do to make the components perform properly.

The Video Camera The video camera is the starting point for the signals that travel the electronic pathway. The camera's job is to convert light and sound into electronic signals and to send those signals to the videotape recorder. Light enters the camera through the camera's lens. By pointing the camera at a scene and adjusting the controls, you can determine just what image and how much light enters the camera. Inside the camera, the light strikes the front of a special tube called a *pickup tube*. Through a special scanning process, the pickup tube changes the light entering the camera into an electronic video signal.

Sound enters the camera through the built-in microphone. Inside the microphone the sound strikes something called a generating element, which changes the sound signal into an electronic signal. With the microphone capturing sounds and the lens capturing light, a portable video camera actually creates two electronic signals at the same time: a signal for sound (the audio signal) and a signal for light (the video signal). During the videotape recording process, these two signals leave the camera and travel side by side through wires to the next point on the electronic pathway—the videotape recorder.

The Videotape Recorder (VTR) The VTR uses very sophisticated electronic machinery to store the video and audio signals as magnetic patterns on a moving videotape. When you play back a videotape, the VTR changes the audio and video information stored on the videotape back into electronic signals again. You could say that a VTR "writes" video and audio information onto the tape during the recording process. When you play back the program, the VTR "reads" that same information from the videotape and sends the electronic signals through a cable to the TV set.

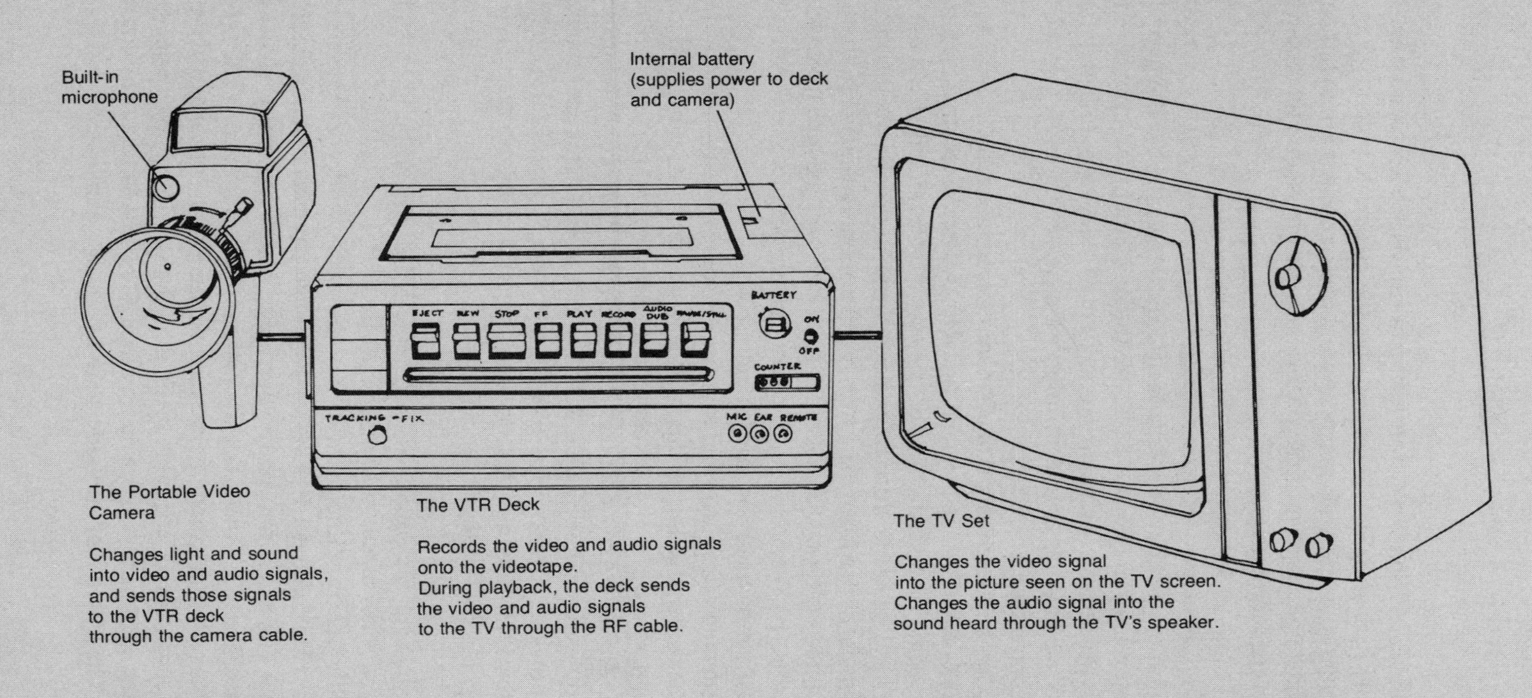

Figure 2.1 *The electronic video pathway.*

The Television Set The television set is the last station on the electronic pathway. The TV set accepts the electronic video and audio signals sent from the VTR and changes those signals back into light and sound. The light appears as the picture on your TV picture tube, and the sound comes through your TV speaker.

The Videotape To record video programming, you will need one more component: the videotape itself. In a way, the videotape is the most important part of a video system. If the videotape is damaged or not working properly, your video production will not record or play back correctly—no matter how carefully you have set up and operated the other components.

Videotapes currently manufactured for home use are called videocassettes, meaning that the tape and the two reels are enclosed in a plastic casing, or cassette. Today, the two most common videocassette formats used for home video recording are the half-inch Beta format and the half-inch VHS format. The inch measurement indicates the width of the tape, and the term "Beta" or "VHS" indicates the type of cassette. Chances are good that the home videocassette recorder you are using is either a Beta or VHS format machine.

Videotapes also vary in length, depending on the amount of tape inside the cassette. A longer tape will give you more recording time, but longer tapes are generally not as durable as shorter-length tapes. This is because the tape must be thinner in order to fit more of it inside a standard-sized cassette. Table 2.1 lists the tape lengths and playing/record times of several popular VHS and Beta videocassettes. As the table shows, the playing and recording time available from each cassette also depends on the tape speed you use.

When you are buying or borrowing a videocassette, check to make sure the cassette fits your deck. Beta tapes will not fit properly into VHS decks and VHS tapes will not fit into Beta decks (Fig. 2.2).

Although it is important to buy the correct videocassette format (Beta or VHS), you don't necessarily have to buy a tape that is made by the same company that made your deck. There are many videotape manufacturers who specialize in making both Beta and VHS tapes. TDK, Maxell, Fuji, Scotch, and several other companies make some of the finest videotapes available, even though these firms don't manufacture VTR machines. When you buy tape, it is best to get a well-known brand. The quality of the tape may affect the quality of your video

Table 2.1 Popular VHS and Beta Videocassette Lengths

VHS Cassettes	*Approximate Length of Tape Inside Cassette*	*Recording/Playback Time At:* Standard Play (SP) (1.32 inches per second)	Long Play (LP) (.66 inches per second)	Super Long Play (SLP) (.45 inches per second)
T-30	225 feet	30 minutes	1 hour	1½ hours
T-60	420 feet	1 hour	2 hours	3 hours
T-120	815 feet	2 hours	4 hours	6 hours
T-160	1080 feet	2 hours 40 minutes	5 hours 20 minutes	8 hours

Beta Cassettes	*Approximate Length of Tape Inside Cassette*	*Recording/Playback Time At:* Beta I Speed (1.54 inches per second)	Beta II Speed (.79 inches per second)	Beta III Speed (.53 inches per second)
L-250	250 feet	30 minutes	1 hour	1½ hours
L-500	500 feet	1 hour	2 hours	3 hours
L-750	750 feet	1½ hours	3 hours	4½ hours
L-830	830 feet	1 hour 40 minutes	3 hours 20 minutes	5 hours

Figure 2.2 ***Beta vs. VHS cassettes. Make sure you have the type of cassette that fits your deck's tape compartment.***

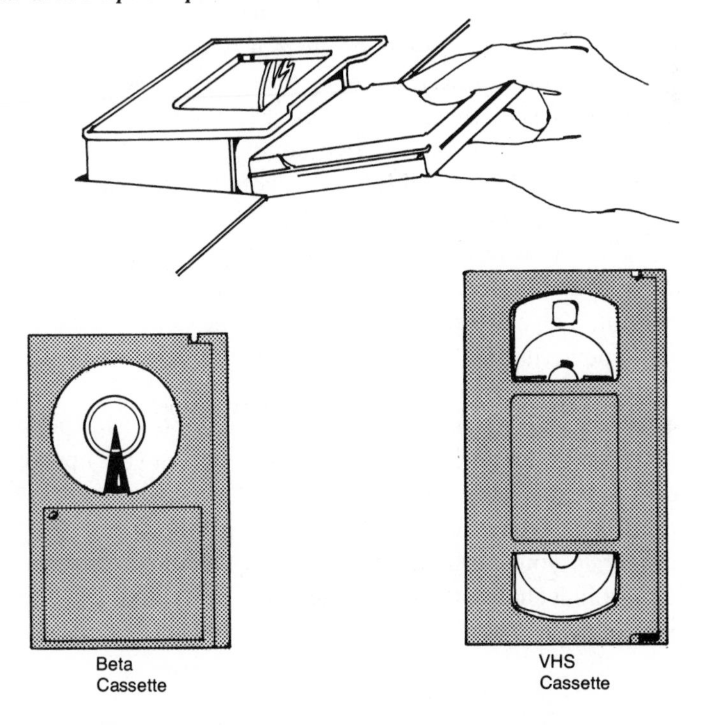

image, and some brands of tape are noticeably better than others.

Here are three pointers for taking care of your videocassettes:

1. When storing cassettes, stand them on end. Do not lay them flat. Laying them flat can damage the edges of the tape, where important information is recorded.
2. When recording or playing a cassette, wind the tape all the way to the end, but don't rewind it. Save the rewinding for the next time you use the cassette. Rewinding a tape just before you use it loosens up the tape so that it moves easily, just as warm-up exercises help athletes loosen-up for a contest.
3. If possible, keep your videocassettes in a cool, dry place. Too much heat, too much cold, or too much moisture can damage the tape.

Feeding Power to the Video System

Before you can begin recording, you'll need to feed electric power into each component of your video system. Electricity is the "fuel" that moves the video and audio signals on their way along the electronic pathway. The electronic power for both the camera and the deck is usually fed into the system through the deck. Most of the time, a VTR's electric power comes from one of two sources: the deck's *internal battery* (all portable VTRs are able to use batteries that fit inside the deck), or the video system's *AC adapter*. The AC adapter allows you to run your deck from an ordinary electrical outlet. Unless your TV is battery-powered (and it probably isn't), you'll have to locate a separate wall socket for it. You can't power a TV set with the same battery or AC adapter that is powering the deck and camera.

AC Adapters and Internal Batteries The circuits inside portable video systems are designed to operate on direct current (DC) power— the kind of power a battery supplies. Power that comes from a wall socket is a different type of current, called alternating current (or AC) power. To operate your portable video system from an ordinary wall outlet, you must first change the wall outlet's AC power into DC power. This is the job of the AC adapter. It converts 110-volt alternating current to 12-volt battery-type direct current, and then feeds the DC power into the deck (Fig. 2.3).

Some manufacturers of portable video equipment offer an AC adapter that is combined with a "tuner." The tuner allows you to use your portable VTR to record your favorite TV programs "off the air." Although the tuner can be a very useful addition to your portable video system, we'll describe the AC adapter as a separate component whose only job is to provide power to the camera and deck.

If you want to use your portable video system in a location where there is no AC power available, you'll have to rely on batteries alone for power. On most portable video systems, the deck's internal battery won't give you more than 45 minutes of recording time before it needs to be recharged. Your operator's manual should indicate how long you can expect the battery to provide power for your particular system. However, if your battery is old or if you are using it in cold weather or if you are using a camera other than the camera that came with your video system, the life of the battery may be much shorter. To be safe, you should figure that your battery will give you about 25 percent less recording time than what the operator's manual claims.

All portable VTRs have a battery meter or battery indicator light that warns you when your battery power is getting low (Fig. 2.4). As soon as the indicator light comes on, or as soon as the needle on the battery meter reaches into the danger zone, you should stop recording. If you ignore the warning signals and continue recording, you'll risk damage to the battery *and* the video recorder. Some decks have special circuitry that turns the system off when the battery power drops below a safe level.

Figure 2.3 ***AC adapter/battery charger***

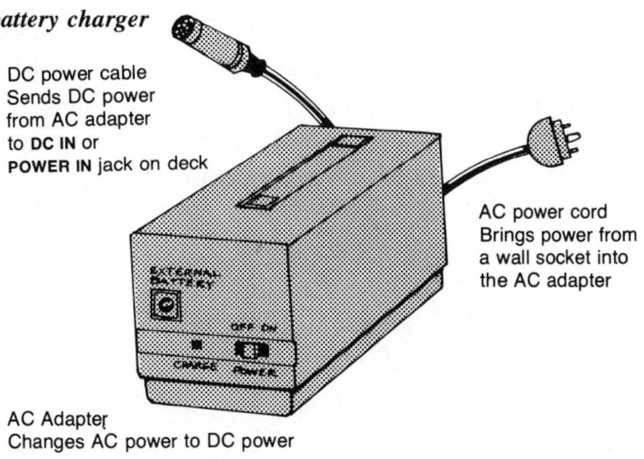

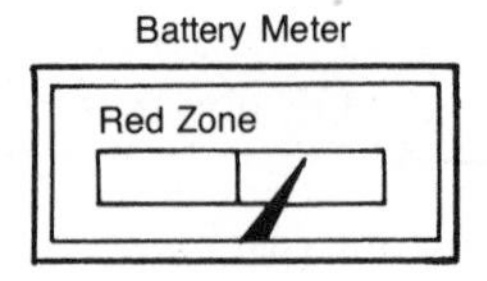

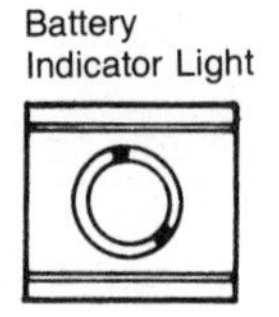

Figure 2.4
A meter or light on the deck will warn you when the battery power is low.

When you need to remove or replace the internal battery, on most portable decks you'll find it in a small compartment behind a door marked "battery." If the door is held on by a screw, loosen the screw and remove the door. Then gently disconnect the plug where the battery is connected to the deck and slide the battery out (Fig. 2.5). On some newer decks, the battery is located in a slot at the back of the deck. On these decks, you simply press the release button near the slot and the battery slides in or out.

Recharging the Battery You can recharge your video system's batteries in either of two ways: You can connect your video system's AC adapter to the VTR and recharge the battery while it remains inside the deck, or you can remove the battery from the deck and connect it to the AC adapter separately (Fig. 2.6). Many of the newer AC adapters have an output jack, or slot, that allows you to make this "outside" battery connection, but don't worry if your AC adapter doesn't. Charging the battery inside the deck works just as well, and it's even a little easier.

To charge the battery while it's inside the deck, connect the cable leading from the AC adapter to the **DC IN** or **EXT POWER IN** jack on the deck. Once you've made this connection, you're ready either to run your VTR on AC power or to recharge the VTR's internal battery. To operate the deck on AC power, turn on the AC adapter's power switch and the VTR power switch (if your VTR has a power switch). Some of the older AC adapters have a special switch that can be set either to **VTR** or to **CHARGING**. Set this switch to **VTR** when you want to operate the portable video system on AC power. But when you want to recharge the deck's battery, set the switch to **CHARGING** instead and turn on the AC adapter.

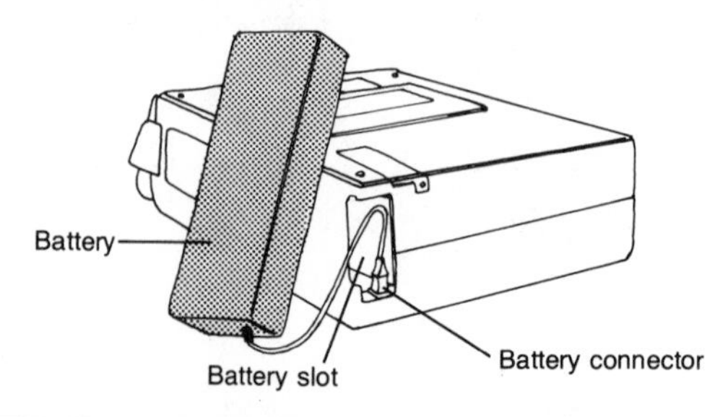

Figure 2.5
Battery slot on a VHS deck, showing battery removed but still connected to deck.

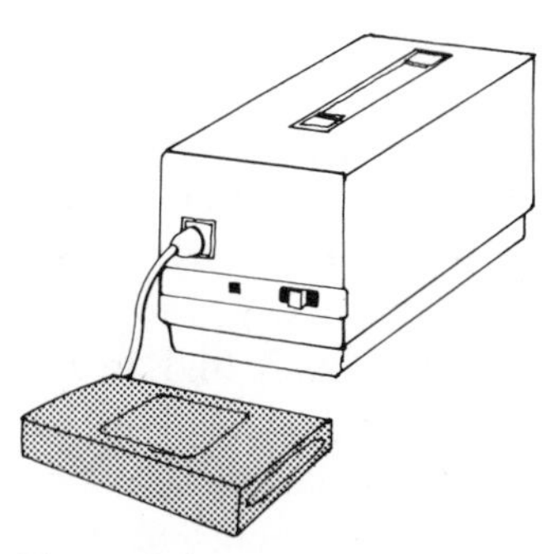

Figure 2.6
Battery removed from deck and connected to AC adapter for recharging.

Figure 2.7
Camera operator wearing a battery belt pack.

When you recharge the deck's battery, you'll want to leave the deck's power switch (if it has one) in the OFF position. This way, all the power will flow into and recharge the battery. When you operate your portable video system on AC power, all of the power coming into the deck goes into running the deck and camera. Since none of the power will reach the battery, the battery won't be recharging.

Most AC adapters have a meter that tells you when your battery is fully charged. You should allow several hours—overnight if possible—for your battery to charge. Don't worry about overcharging the battery. Most AC adapters have circuitry that regulates the charge as the battery becomes stronger. If you are using the battery and AC adapter that were originally provided with the deck, you shouldn't have any problem. (You'll have to be more careful if you're using an AC adapter or battery that didn't come with your video system.)

Other Power Options There are two other power options that are available as add-on features to most portable video systems: *car battery cords* and *external battery packs*. A car battery cord plugs into a car's cigarette lighter and allows you to run your portable video system off the car's 12-volt battery. An external battery pack plugs into the DC INPUT on the deck and gives you an extra source of power once the deck's internal battery runs out. Some external batteries come in a special *belt pack* (Fig. 2.7). As you might guess, you wear the belt pack around your waist. When the power in the external battery pack gets low, you can recharge the pack by plugging it into your AC adapter.

Whenever possible, you should save battery power by using the AC adapter to run your portable video system. Unlike batteries, AC current provides an almost unlimited supply of power. If you're using the equipment outdoors or in a location where you can't find an AC socket to plug into, the portable video system will automatically begin running on battery power once you disconnect the AC adapter. Just be sure that you have a battery in the deck's battery compartment and that it is fully charged. And remember that once you are operating on battery power, you are operating on borrowed time! Plan your recordings carefully, so you have enough battery power to finish your recording session.

Making Good Connections

You now know that the camera, the VTR deck, and the TV set serve as the three major stations on the electronic video recording/playback pathway. In this section, you'll learn about the inputs, outputs, and cables that allow you to connect these three major components. With the right cables and connectors, you'll be able to connect the points on the electronic pathway so the audio and video signals go where they're supposed to go. Without the correct cables and connectors, the signals won't flow properly, and you won't be able to record and play back video productions.

Inputs and Outputs Before trying to connect any components, you should become familiar with inputs and outputs (or jacks) on your VTR deck. As you know, the VTR deck is right in the middle of the electronic pathway, coming between the camera and the TV set. The most important jacks on the deck are the *inputs*, through which the deck receives audio and video signals from the camera, and the *outputs*, through which the deck sends signals to the TV set (Fig. 2.8). Most decks also have several other inputs and outputs. You're already familiar with one of these inputs—the DC IN or EXT POWER IN input that allows you to send power from the AC adapter into the deck. Check the front, back, and side panels of your deck to see which of the following inputs and outputs are featured on your portable video system.

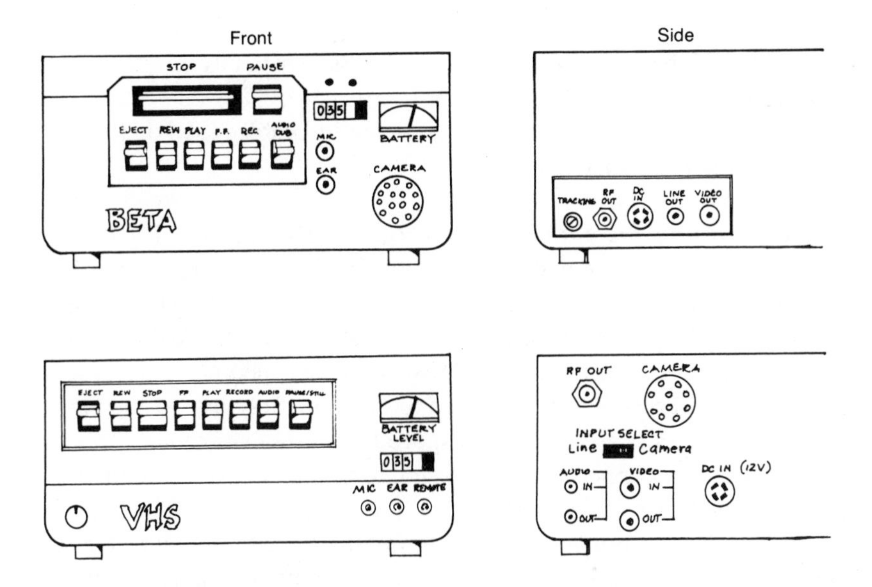

Figure 2.8 ***Sample front and side panel views of VHS and Beta decks, showing typical inputs and outputs.***

INPUTS

- **CAMERA IN**
- **AUDIO IN**
- **TV** or **VHF IN**
- **DC,** or **EXT POWER IN**
- **MICROPHONE,** or **MIC IN**
- **VIDEO IN**
- **AUDIO IN** or **LINE IN**

OUTPUTS

- **VIDEO OUT**
- **AUDIO OUT**
- **TV, VHF,** or **RF OUT**

On all decks, *inputs* accept signals coming into the deck from other components, and *outputs* send signals out from the deck to another component. These signals are carried through cables, which you use to connect components together. At the end of a cable, you'll usually find a *plug*. Plugs come in a confusing array of shapes and sizes. To connect a cable to a piece of equipment, the plug must fit its corresponding jack. You can't fit one kind of plug into a different kind of jack.

Most portable video systems are sold with all the cables you need to connect the components for video recording and playback. If your system doesn't have all the right cables, you'll have to buy or borrow the missing ones.

Figure 2.9 includes some typical plugs and cables used on today's portable video systems.

Connecting the Camera to the Deck For the first leg of the electronic pathway, the audio and video signals travel through a cable that connects the video camera to the VTR deck. Connecting the camera to the VTR is really very simple, as long as the camera and deck are made by the same manufacturer. Most portable video cameras have a camera cable permanently connected to the camera body. At the other end of this camera cable there is a special multi-pin connector. Take a good look at this connector before trying to attach it to the deck. If your camera is part of a VHS system, you'll see that the connector has ten small pins. Not surprisingly, this plug is called a *ten-pin connector*. If the camera you are using is part of a Beta system, you'll see fourteen pins inside the connector. This is called a *fourteen-pin connector* (Fig. 2.10).

The VTR's camera-input jack is usually located on one of the deck's front or side panels. The number of holes in the deck's camera-input jack should equal the number of pins in your camera-cable connector. You should also find a notch at the top of the deck's camera-input jack that matches a slot in the camera-cable connector (Fig. 2.11). To connect the camera cable to the deck, align the slot in the ten-pin or fourteen-pin connector with the notch on the camera-input jack. Then push the connector gently but firmly into place. Many camera-cable connectors have a threaded metal sleeve that prevents the camera cable from being accidentally disconnected during recording. After sliding the camera-cable connector into the camera input jack, tighten this sleeve until it is snug.

To connect a portable video camera and portable video deck made by different manufacturers, you may have to buy a special adapter. Beta and VHS format decks use different camera-cable connectors, so it's impossible to connect a Beta camera directly to a VHS deck, or a VHS camera directly to a Beta deck.

If you want to connect two different brands of components within the same format (say a Panasonic camera made for a VHS system to a Hitachi VHS deck), you may still run into trouble. Different VHS manufacturers sometimes use different wiring systems for the pins in their ten-pin connectors. This means that even though the pins may fit

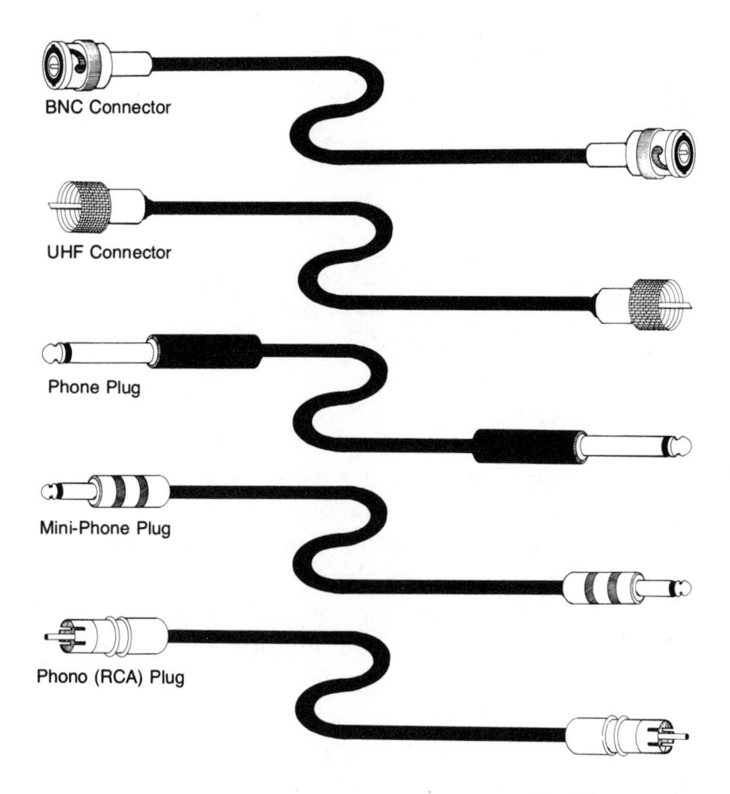

Figure 2.9 *Some typical plugs and cables used in portable video systems.*

into the holes on the jack, the right pins may not be in the right holes. If this happens, signals will not flow properly between the camera and the deck. Adapters that will allow you to connect a camera and deck made by different manufacturers are available at many video stores, but they can be expensive ($80–$100 each).

There is one very useful cable that may not have come with your portable video system. This is a camera extension cable. They come in several lengths and are fairly expensive ($90–$120), but they are well worth the extra price. With a 30-foot camera extension cable, you are able to move your camera quite freely during a shooting session without having to move the deck around at the same time.

Connecting the VTR to the TV As you know, when you play back a videotape, the deck "reads" the magnetic signals stored on the tape and sends the signals on to a television set through one of its output jacks. There are two types of output jacks on most VTRs. One type is for connecting the VTR to other components, such as another VTR or a special *TV monitor*. The other type is for connecting the deck to a regular

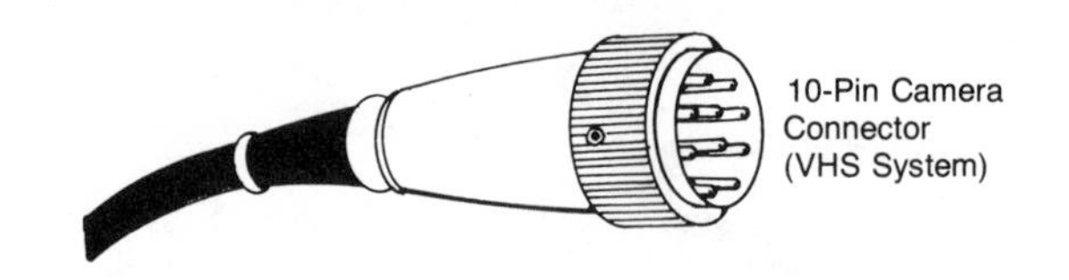

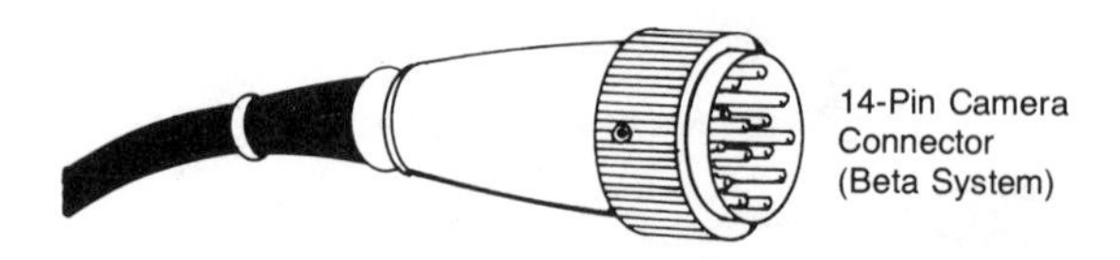

Figure 2.10

home-type *TV receiver*. A TV monitor cannot receive ordinary television broadcast signals. It can only receive video and audio signals sent directly from a VTR, a video camera, or some other component that is sending out a video signal. TV monitors are usually used in professional situations, not in homes. A TV receiver, on the other hand, is a TV set designed to pick up ordinary broadcast channels in the home. Therefore, a TV receiver has a channel selector knob (or buttons) so that you can watch programs broadcast from television stations.

Since you aren't likely to have a special TV monitor, we'll explain how to connect your VTR to an ordinary home-type TV set. First, locate the output jack marked TV OUT, VHF OUT, or RF OUT on the side panel of the VTR. This output jack will have a very small hole in the center and will probably have metal threads around the outside. The type of connector that fits into this jack is called an RF connector or an F-type connector. It has a very thin pin in the center and a threaded metal collar on the outside that tightens around the jack (Fig. 2.12). (Some of the newer VTRs use RF connectors that slip on rather than screw on.)

The cable that carries signals to the TV set is called an RF cable. First, connect the RF connector on one end of the RF cable to the VHF or RF OUT jack on the deck. Then turn the TV set around so that the back side of the set is facing you. Some of the newer TV sets have a jack marked VHF IN or RF IN on the back panel so you can connect an RF cable directly to the set. Most TV sets, though, require that you connect the RF cable coming from the deck to the VHF antenna terminals on the back of the set. To do this, you'll need an adapter, or "matching transformer" (also called a 75-300-ohm converter) that has an RF jack at one end and two prongs at the other (Fig. 2.13). Push the prongs onto the terminals marked VHF and tighten the screws to hold them on. Connect the RF cable to the other end of the matching transformer, and you've completed the "VTR-TV" connection.

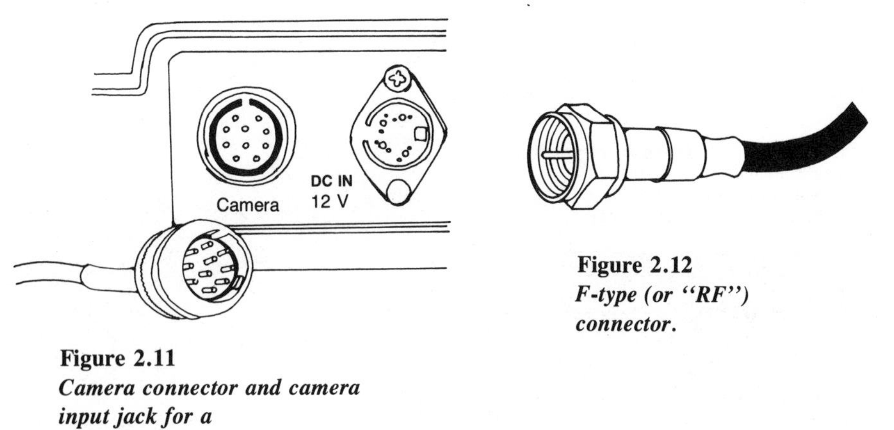

Figure 2.11
Camera connector and camera input jack for a VHS portable video system.

Figure 2.12
F-type (or "RF") connector.

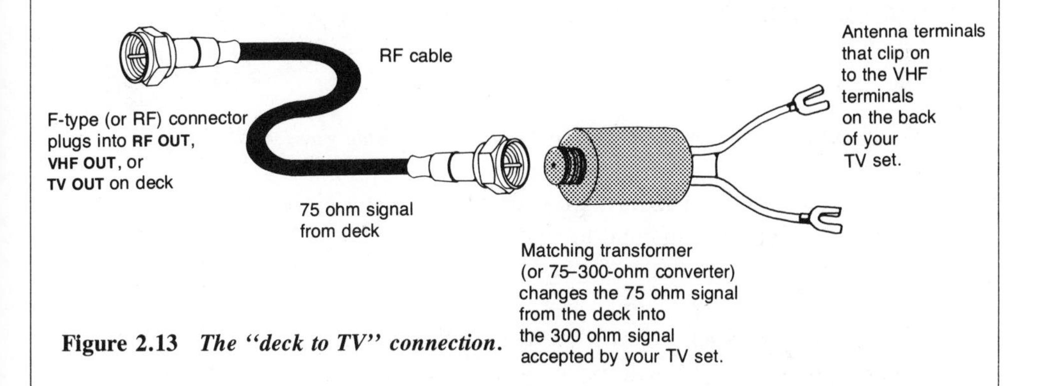

Figure 2.13 *The "deck to TV" connection.*

Actually, there's one more step you should know about. The signal that is fed into the TV set from the VTR is not exactly the same type of signal that first entered the VTR from the camera. Instead, the signal that is sent from the **RF OUT** or **VHF OUT** jack on the deck is like the regular TV broadcast signals you receive through your TV set. Inside the VTR, a device called an RF converter changes the video and audio signals into an RF signal. This RF signal (or radio frequency signal), matches one of the channels of your TV set.

Most VTRs put out an RF signal that matches Channels 3 or 4, and many decks have a switch that allows you to choose between Channels 3 and 4. If your deck has this channel selector switch, find out which channel is *not* used by a TV station near where you live and leave the switch on the deck in that position. Then, when you play back the tape through a TV set, make sure the channel selector on the TV is tuned to the same channel the deck is sending to it (Fig. 2.14). That way, the deck and the TV set will be sending and receiving on the same wave length.

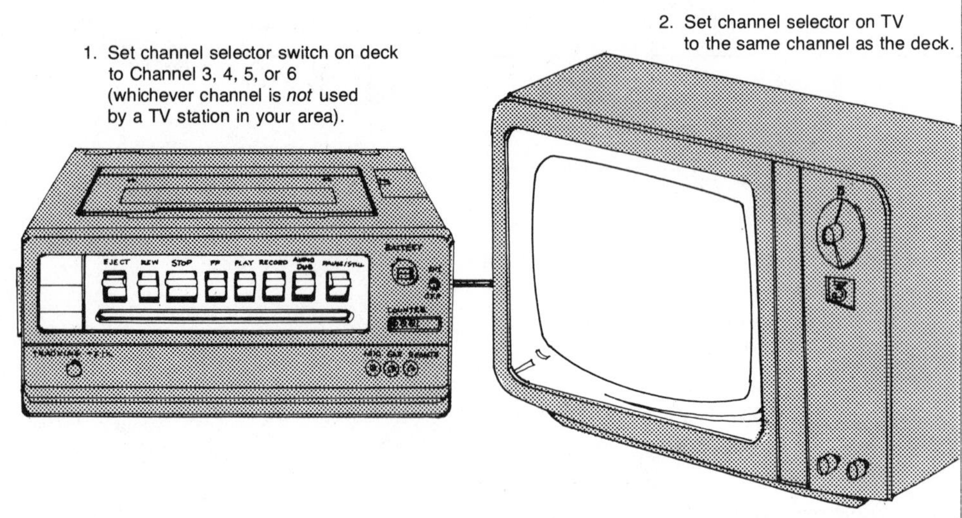

Figure 2.14 ***Make sure the deck and TV are on the same wavelength.***

(If your TV is hooked up through your videotape deck's *tuner/timer* box, read your operator's manual to find out the best way of connecting the deck to the TV.)

One Last Step If you've been following along, your portable video camera should now be connected to your deck, and the deck should be connected to a TV set. The deck should also be connected to an internal battery, an AC adapter, or one of the other power sources we described. If you've made these connections, you're ready for one last step—loading the cassette in your VTR.

First, press the EJECT button on the deck. (If your deck has a power switch, you'll have to turn it on first.) When the tape compartment pops open, turn the videocassette so that the side where the tape is exposed faces into the deck, and so the holes of the tape reels are facing down. Slide the cassette gently into the tape compartment and push the compartment back down into the machine. With the videocassette in the deck, you should be all set to start video recording.

chapter 3

GETTING STARTED: MAKING YOUR PORTABLE VIDEO SYSTEM WORK FOR YOU

Once you've connected the pieces of a portable video system, you're ready to start your video system working for you. In this chapter, you'll learn what the different lights and buttons on your VTR deck do and what buttons to push to record and play back video productions. You'll also learn how to set up and operate your portable video camera and how to use your camera's built-in microphone. Finally, you'll find out how to take care of your portable video system and how to find and fix simple equipment problems.

Pushing the Right Buttons

All portable VTRs have a row of buttons (or function controls) located in front of the tape cassette compartment (Fig. 3.1). Some decks have more function controls than others. Generally, these controls have the following labels:

- **EJECT**
- **REC** (record)
- **PLAY** (or **FORWARD**)
- **PAUSE**
- **STOP**
- **REW** (rewind)
- **FF** (fast forward)
- **AUDIO DUB**
- **SCAN** (or **SEARCH** or **SHUTTLE SEARCH**)

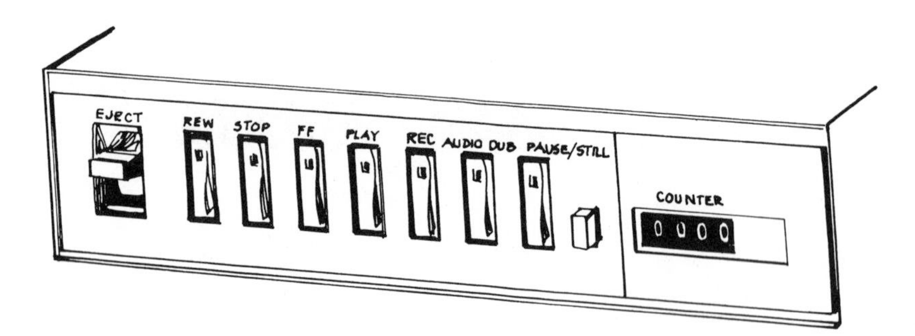

Figure 3.1 ***Function control panel on a typical portable video deck.***

- EJECT opens the tape compartment so you can load or remove a cassette.
- REC (record) turns on the electronic circuits that record video and audio signals onto the videotape. On most portable decks, you must push PLAY and RECORD at the same time to begin recording on a tape. (Pushing PLAY gets the tape moving, and pushing RECORD starts the electronic recording process.) Some of the newer decks will begin recording with only a push of the RECORD button. Once you put the deck into the RECORD mode, you can start and stop the tape by squeezing the portable video camera's trigger.
- PLAY is the most important control on your deck. Pushing PLAY moves the tape forward at the correct speed for playback and recording and switches on the electronic circuits that "read" the video and audio signals on the tape so that they may be sent to a TV set. On some decks, the "play" function is called FORWARD, or FWD.
- PAUSE is a special tape function that you are most likely to use when the deck is in PLAY or RECORD. When you push the PAUSE control during playback, the tape stops moving but the deck continues sending a video signal to the TV set. The result is a *freeze-frame effect*—the image on the TV screen "freezes" and the sound stops. When you push the PAUSE button a second time, the tape will start moving again and the sound will return. As you will learn in Chapter 4, you can also use the PAUSE function to piece together (or edit) different shots and scenes into a single videotape.

  A word of caution—you shouldn't leave your deck in PAUSE for more than a few minutes. Leaving the deck in PAUSE for longer periods can wear out the videotape and cause damage to the deck's video recording heads. For recording breaks of more than two minutes, use the deck's STOP control to stop recording. Some decks will automatically switch to STOP if you leave them in PAUSE for too long.
- STOP control stops everything. When you push STOP, any other tape control that you might have activated will release, the recording circuits will shut down, and the tape will stop moving. When you go from PLAY to REWIND or FAST FORWARD, it's a good idea to push STOP first. On some decks, especially older

decks, going from one function to another without pushing STOP can cause damage to the tape and deck.

- REW (rewind) control winds the tape backward at a high speed so you can play back or record over an earlier section of tape. Unless your deck has a reverse SCAN or SEARCH feature (see below), it will show only snow (static) on a TV screen when you rewind a tape.
- FF (fast forward) is just the opposite of REWIND. FAST FORWARD winds the tape quickly ahead, in the same direction as PLAY. Use FAST FORWARD when you want to skip over a section of the tape so you can play back or record over a section that is farther ahead.
- SCAN or SEARCH Some newer decks have a SCAN or SEARCH feature that works much like FAST FORWARD. There is one important difference though. With SCAN or SEARCH, the picture stays on the TV screen while the tape is rapidly moving forward. (Some decks also feature a REWIND SCAN function.) The tape does not move quite as quickly as it does in FAST FORWARD, but you are able to see all of the images on the tape while it is moving at high speed. By watching the TV screen as you use the SCAN or SEARCH control, you can move the tape forward and backward to find the exact scene or section you want.
- AUDIO DUB If your deck has an AUDIO DUB control, you can use it to record a new audio signal without changing the video signal. You'll learn how to use AUDIO DUB, and what you might use it for, in Chapter 4.

Other Lights and Switches

In addition to the deck's basic function controls, there are several other lights and switches you should become familiar with before you begin to operate your video equipment. These lights and switches are listed on page 43. Check to see which of the listed items are featured on your deck. Learn what any warning lights mean, and pay attention to the lights as you record and play back videotapes. You should also consult your operator's manual to find out how you use any unfamiliar lights and controls during video recording and playback.

LIGHTS/METERS

- dew indicator
- battery level meter
- standby lamp (Beta decks only)
- pause indicator

SWITCHES

- speed control
- tape counter reset switch
- memory rewind switch
- video dub
- record lock
- tracking control

The speed control switch on Beta and VHS decks deserves special mention. Newer portable decks offer a choice of recording speeds, so you can get more recording time out of one videocassette. (Until recently, only larger home console VTRs had more than one recording speed, so don't be surprised if your portable deck has no switch marked SPEED or RECORD SPEED.) If you have a VHS deck with variable recording speeds, these different speeds will probably be labeled SP (standard play), LP (long play), and SLP (super long play). Using a standard T-120 VHS cassette, the SP speed will give you two hours of recording time, LP will give you four hours, and SLP will give you six hours. You can get the longer recording time because the tape moves more slowly at the LP and SLP speeds.

On Beta decks, the different recording speeds are indicated as X-1, X-2, and X-3. The earliest Beta portable decks recorded and played back only at the X-1 speed (one hour of recording time on a standard L-500 Beta cassette). The newer Beta portables either operate at only the X-2 speed (two hours of recording time on an L-500 Beta tape), or let you choose between the X-2 and X-3 speeds (X-3 gives 3 hours of recording time on an L-500 Beta tape).

As a general rule, you should always operate a portable deck at the standard one- or two-hour speed—the fastest tape speed the deck uses. When you record at the slower speeds, your tape will become "packed" or saturated with too much video information. As a result, the image that you see during playback will not be as good as it would be if you had recorded at the standard speed. If you're using batteries to power your deck, you won't be able to record for much longer than an hour anyway. The batteries won't last that long!

The last feature that you should know about before you record or play back a videotape is the deck's tape counter (Fig. 3.2). When you're ready to record, push the counter RESET button (located right next to the tape counter on most decks). This will set the counter at "000." When you are ready to view the segment that you just recorded, you can find

Figure 3.2 ***Counter set at "000" for the beginning of a recording session.***

the beginning of the segment by simply rewinding the tape until the counter reads "000" again. You may want to keep a pad and pencil nearby while you record and jot down the number on the counter whenever you begin to record a new segment. This will allow you to find each segment quickly when you return to view the tape.

One final word about tape counters: Unless it says so right on the deck (and it usually doesn't), a tape counter will not measure minutes or seconds. Most tape counters measure the number of revolutions of the take-up reel inside the cassette. Some decks come with a chart that allows you to translate the counter reading into minutes and seconds. Be aware, however, that these charts are not completely accurate. A watch with a sweep secondhand or a digital stopwatch gives you a more accurate way to time your productions.

Using the Color Camera

In the early days of color television, color camera operators had to be very patient and painstaking people. In those days, color cameras were clumsy and heavy, and the camera's color tubes and circuitry needed constant attention and adjustment. With all this to worry about, many early color camera operators must have wondered if color television was really worth the extra effort!

Fortunately, today's color cameras are much easier to operate. Those made for home use are usually small and lightweight, and many have controls that automatically adjust to different color and lighting conditions. To use your portable video camera, you'll first need to know about *camera viewfinders*.

Camera Viewfinders You use your camera's viewfinder to set up and *frame* shots during a video production. There are three basic types of video camera viewfinders:

1. optical viewfinders
2. through-the-lens viewfinders
3. electronic viewfinders

- *Optical viewfinders* are "peep" lenses mounted on the side or top of the camera, much like the viewfinders on inexpensive pocket snapshot cameras (Fig. 3.3). Optical viewfinders are the least expensive type and the most difficult to use properly. Because they are mounted above or to the side of the camera lens, optical viewfinders do not allow you to see what the camera lens sees. This makes it almost impossible to line up and focus shots and to use the zoom lens, unless the camera and deck are connected to a TV set (so you can see the shot the deck is actually recording).
- *Through-the-lens* (TTL) *viewfinders* allow you to look through the lens so you can see what the lens is actually seeing (Fig. 3.4). TTL viewfinders are more expensive than optical viewfinders, but they make it much easier to focus the camera and use the zoom lens. With a TTL viewfinder, you can be sure that the image you are seeing is the image you are recording on the videotape.
- *Electronic viewfinders* are the most expensive, but they offer one important advantage. Because they include a tiny TV screen, electronic viewfinders allow you to see the same video signal that is being sent to the deck (Fig. 3.5). You can also use the electronic viewfinder screen to play back a segment of videotape you've just recorded, so you can review and check the picture quality of the segment. (Since most electronic viewfinders have black-and-white screens, you probably won't be able to check the color quality.) Electronic viewfinders are especially handy when you are recording outdoors, where you may not have access to a TV set for reviewing segments of your tape.

Check to see which type of viewfinder your camera has. If you are using an optical viewfinder, remember to check your shots on a TV screen. Otherwise, you may find you've recorded shots that are out of focus, or shots that cut off (or crop) people or objects you meant to include.

Camera Lens Control Rings Most cameras have three control rings located on the lens barrel: the aperture ring, the zoom ring, and the focus ring (Fig. 3.6). These rings are the most important controls on the camera. You use them to select, focus, and add emphasis to the scenes and images you are recording.

Figure 3.3 ***Portable video camera with optical (peep-type) viewfinder.***

Figure 3.4 ***Portable video camera with through-the-lens (TTL) viewfinder.***

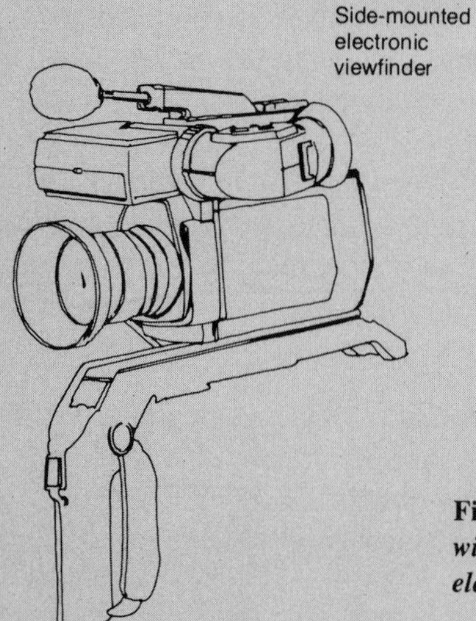

Figure 3.5 ***Portable video camera with side-mounted electronic viewfinder.***

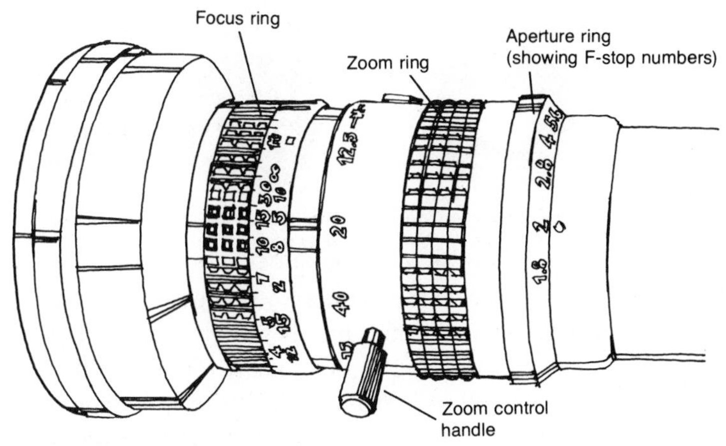

Figure 3.6 *Controls rings on a video zoom lens.*

The *aperture ring* controls the amount of light entering the camera. When you turn the ring, you adjust the metal leaves that make up the lens iris. Turn the aperture ring one way and the metal leaves open up to create a larger iris opening. This lets more light into the camera. Turn the aperture ring the other way and the metal leaves close in to create a smaller iris opening. This lets less light into the camera.

Many newer color cameras have a control that opens and closes the iris automatically. Some color cameras that have this *automatic iris control* do not have an aperture ring. Check for an aperture ring on your camera. If your camera has one, it should be the control ring closest to the body of the camera (Fig. 3.6). If your camera has an automatic iris control and an aperture ring, it probably has a switch on the side of the camera that allows you to override the automatic iris, too. By turning this switch to the MANUAL position, you will turn off the automatic iris. This will allow you to adjust the iris yourself, either by turning the aperture ring or by turning the manual iris control knob on the side of the camera. If your camera has an automatic iris but no override switch, the iris is completely controlled by the automatic iris feature—and it's completely out of your hands.

If your camera has an aperture ring, you'll see several numbers marked on the ring. These are called F-stop numbers, and they usually range from F/16 to F/1.8. It's a little confusing, but the lowest F-stop indicates the widest iris opening, and the highest F-stop indicates the smallest iris opening. In other words, when you set the F-stop at the lowest number (usually F/1.8), the iris opening is very wide and a great deal of light enters the camera (Fig. 3.7). Each higher F-stop number allows less light into the camera. In fact, each time you turn the aperture ring to the next highest F-stop number, you cut the amount of light entering the camera exactly in half!

With the aperture ring set at F/1.8

The iris is fully open, and the maximum amount of light enters the camera.

With the aperture ring set at F/5.6

The iris is partially closed, and less light enters the camera.

With the aperture ring set at F/16

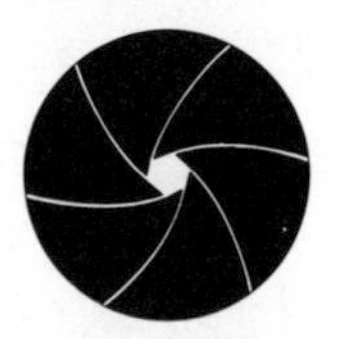

The iris is almost completely closed, and very little light enters the camera.

Figure 3.7
Iris openings and F-stop settings.

When you are setting up for a video recording session, it's not hard to find the correct F-stop setting. Just before you're ready to begin recording, turn the aperture ring to the highest F-stop setting (the one that closes the iris as much as possible). Do this before you remove the camera's lens cap to prevent costly damage to the camera caused by too much light entering the lens too quickly. For the same reason, *you should never point a video camera directly at the sun or any bright object*.

Once you've checked to make sure the aperture ring is at the highest F-stop setting, remove the lens cap and aim the camera at the scene you plan to record. Look into your camera's viewfinder and slowly turn the aperture ring until an image appears. As you turn the aperture ring, you'll feel each F-stop "click." Turn the aperture ring until the image in the viewfinder looks bright and the shadows and bright spots in the scene look natural and balanced. To get a clear image in the viewfinder, you may also have to adjust the zoom and focus controls described later. (If your camera has an optical viewfinder, you will have to use a television screen to adjust the aperture ring to the proper setting. The optical viewfinder screen will not show you what the camera is actually seeing.)

For most recording locations, an aperture setting between F/2 and F/8 should produce a balanced image in your viewfinder. If you set the aperture ring at too high an F-stop number, you will see a dim washed-out image on your viewfinder or television screen. An F-stop number that is too low will let too much light into the camera, and you'll see nothing but the brightest and darkest parts of the picture.

Beyond the F-stop marked 16, some lenses have a *C* setting. When set at *C*, a lens iris is fully closed, allowing no light at all to enter the camera. You can use the *C* setting to protect the camera tube when the camera is not in use. As you'll learn in Chapter 4, the *C* setting also lets you perform some simple "video transitions."

The *zoom ring* is the middle control ring on the camera lens barrel (Fig. 3.6). By turning the zoom ring, you can make a scene seem closer or farther away from the camera than it appears to your naked eye. Like the aperture ring, the zoom ring is marked with a series of numbers. By turning the zoom control to the highest number marked on the ring, you'll bring people and objects "close" to the camera. By turning the zoom control to the lowest number marked on the ring, you'll make people and objects seem to move away from the camera.

When the zoom ring is set at the CLOSE-IN position (the highest zoom ring number), the lens is said to be at the *maximum telephoto setting*. Turning the ring toward this setting is called "zooming in." When you zoom in on a subject (a friend for instance), the subject appears to grow larger, and much of the background scene disappears from the frame. In other words, when you zoom in to the maximum telephoto setting, you make your subject appear to stand out from the rest of the scene. Using the zoom lens at the maximum telephoto setting is almost like using a telescope. You get a close-up of the main part of a scene, but you miss much of the action around the edges of the frame.

Some newer video cameras have a motorized zoom control—or *power zoom* as it is sometimes called. If your camera has a power zoom feature, you operate the zoom by pushing a button on the top or the side of the camera. A power zoom can help you to zoom in and out smoothly, but it can also make it hard for you to control the speed of the zoom. For more about the way your zoom ring works, see the Beat the Zoom game at the end of this chapter.

The *focus ring* is the ring farthest out on the lens barrel (Fig. 3.6). When the focus ring is properly adjusted, the image in your camera viewfinder appears sharp and clear, and the camera is said to be "in focus." Focusing the camera isn't very difficult. If your camera has an electronic or a through-the-lens viewfinder, you simply turn the focus ring until the image in your viewfinder appears sharp. To guarantee good focus with an optical viewfinder, you'll have to watch the image on a TV screen as you adjust the focus ring.

Some video cameras have an automatic focus that automatically adjusts the focus as you use your zoom ring. If your camera doesn't have this feature, you may find it hard to keep the image in focus as you zoom in or out on a subject. If you run into this problem, try following these simple steps:

1. locate the subject in the center of the viewfinder
2. zoom all the way in on the subject (to the maximum telephoto setting)
3. adjust the focus ring until the subject is sharp and clear
4. zoom out to the desired closeness and framing for the shot and begin shooting

Follow these focusing steps every time you set up a shot, even if you don't plan to zoom in or out during the shot. If you've followed each step, your image will stay in focus throughout the zoom range—unless you or your subject moves. If either of you *does* move, repeat the steps.

Getting the Camera to Show Its True Colors Unlike the human eye, most color video cameras don't automatically adjust to changing light conditions. Different types of light affect the camera in different ways. For example, if the camera is set for sunlight, using the camera in a room with fluorescent (tubelike) lights will add a greenish tint to the video recording. On the other hand, normal daylight often has a blue tint when seen through the camera lens, and early morning or late afternoon sunlight is reddish. To make sure that the colors on your videotape look right (with no purple hair or green faces), you'll need to adjust the COLOR TEMPERATURE and WHITE BALANCE controls found on most portable video cameras (Fig. 3.8).

On many portable video cameras, you adjust color tint (or color temperature) by moving a switch that slides special filters behind the lens. This switch is usually marked FILTER or COLOR TEMP, and it usually has settings for daylight (sunlight), cloudy daylight, nighttime, and artificial light (indoor light). Before you begin recording, turn the switch to the setting that matches the type of light at your location.

Most color cameras also need their color circuits adjusted electronically. Your color camera creates a color video signal by combining three primary colors: red, green, and blue. When all three colors are perfectly balanced, they create pure white. Therefore, when a pure white image looks right to the camera (when white is "balanced") you can be sure that the camera's three primary colors will also be balanced.

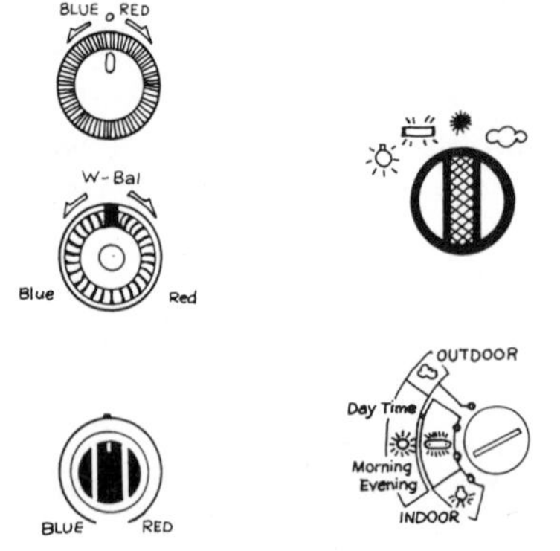

White balance controls

Color temperature controls

Figure 3.8 ***White balance and color temperature controls.***

To balance the white signal on most cameras, you adjust a special camera control. Thanks to the good sense of some long-forgotten video equipment designer, this control is called the **WHITE BALANCE CONTROL**. On most cameras, this control adjusts two of the primary colors (red and blue) in relation to the other constant primary color (green). Read your operator's manual to find out how to operate this control on your camera. Usually, you place a white card (a blank index card will do) a few inches in front of the camera lens and then adjust the **WHITE BALANCE KNOB** until a meter or indicator light tells you the camera is in balance.

Many color cameras also have features that help you adjust for extreme or unusual lighting conditions. One such feature is the **LOW LIGHT COMPENSATION (LLC) SWITCH**. This switch boosts the camera's overall sensitivity to light for situations where there isn't enough light to record a good image. However, we should warn you about the **LLC** control. In addition to boosting the camera's sensitivity to light, the **LLC** also boosts the amount of distortion in the picture you're recording. If possible, you should compensate for a low light level by bringing in extra lights—not by using the **LLC** switch. (We'll explain how to use lights effectively in Chapter 4.)

Some color video cameras also have a **BACK LIGHT COMPENSATION (BLC) SWITCH**. You use this switch when there is a bright light source behind your subject, shining into the camera.

Ready to Record? Now that you know how to set up and operate your portable video system, you're ready for the easy part—actually making a video recording. Here's a 10-step checklist you can follow to make sure you're ready to record. If you run into any trouble, turn back to the pages listed for each step.

1. Make sure the components are connected properly (see Chapter 2).
2. If you're recording near a wall outlet, plug in the AC adapter and connect the adapter to the deck (see pp. 28–31).
3. If the deck has a power switch, turn it on.
4. Load a cassette into the deck. Be sure to rewind the tape now if you want to start recording at the beginning of the cassette (see page 38).
5. Adjust the camera's **WHITE BALANCE** and **COLOR TEMPERATURE** controls (see pp. 50–51). With some portable cameras, you'll have to put the deck in **RECORD** to adjust these controls.

6. Adjust the camera's lens controls until you have a clear focused image on the viewfinder or TV screen (see pp. 44–50). *Be sure not to point the camera at any bright lights.*
7. Push the PLAY and RECORD controls at the same time, (or just RECORD, depending on the type of VTR deck you have) (see pp. 41–42).
8. Squeeze and release the camera trigger. You should now be recording a video production!
9. To "pause" during recording, squeeze and release the camera trigger again. To stop recording, press the STOP control on the deck.
10. Record and play back a short test segment.

A test segment is an especially good idea before you begin recording something important. By watching the test segment, you'll be able to tell whether the components are connected properly, whether the equipment is working right, and whether the camera was focused correctly. If something is wrong, it's better to find out *before* an important production session, when there is still time to correct any problems.

Playing It Back: Using the Deck's Playback Controls

To play back a videotape, you connect your deck to a TV set, tune the TV to the correct channel, insert and rewind the videocassette, and press the deck's PLAY control. If the color balance or picture quality of your playback doesn't seem right, check the controls on your TV set.

As you watch your video recording, you may also want to use the deck's special playback controls to change to fast or slow motion, or to adjust the video image displayed on your TV screen. Check the front and side panels of your video deck to see which of the following playback controls are available on the video system you are using.

- **TRACKING CONTROL**
- **SLOW MOTION**
- **SCAN**
- **FRAME-BY-FRAME ADVANCE** (sometimes found on the hand-held remote-control unit that comes with some portable decks)

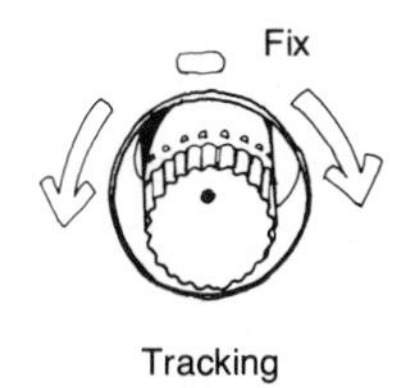

Figure 3.9
Tracking control knob in the normal (or fixed) position.

As a general rule, newer portable video decks tend to have a larger assortment of playback controls than older decks. In fact, if you are using one of the newest portable video recorders, your deck may have playback controls that weren't available on portable decks when this book was being written!

Almost all portable video decks have a TRACKING CONTROL that adjusts the position of the videotape as it passes over the video heads (Fig. 3.9). When you are recording a tape, make sure this control is in the middle, or fixed, position. If it isn't, you may run into a tracking problem the next time you record a tape.

When you play back a videotape, you may occasionally see a horizontal line of distortion (called a tear) cutting straight across the image displayed on the TV screen (Fig. 3.10). If this problem doesn't appear, leave the TRACKING CONTROL in the fixed position. If the distortion does appear, turn the control until the tear disappears. After viewing the tape, be sure to return the control to the fixed position.

If you watch television, you must be familiar with "fast motion" and "slow motion." You've probably seen slow motion used during instant replays on sports programs, and you may have seen fast motion used during comic scenes on cartoons and other entertainment programs. Some newer portable video decks feature SLOW MOTION and SCAN controls that allow you to create similar effects as you play back a tape. If these controls are available on your deck, check your operator's

Figure 3.10 *Correcting a tracking error.*

1. If the picture looks normal during playback leave the deck's

tracking control knob in the normal (or fixed) position.

2. If a tracking error (or "tear") appears on the screen

turn the tracking control knob until . . .

3. The line of distortion disappears and the picture looks normal again.

When the tape is finished, return the tracking knob to the normal position.

manual to see how they work. On many decks, you first put the tape into the PLAY mode and then press the SLOW MOTION or SCAN controls. Pressing SLOW MOTION slows the action below its normal speed, and pressing SCAN speeds up the action to fast motion. On some decks, you can also put the tape into REWIND, and then press SLOW MOTION or SCAN to create slow or fast motion action in reverse. Your deck may also have a 2X control that allows you to play back tapes at twice their normal speed.

Basic Audio: Using the Built-In Microphone

So far, we've been concentrating on the video, or picture, part of your productions. We haven't said much about recording or playing back sound. Inexperienced video producers sometimes seem to forget that the audio signal is as important to the program as the video signal. To create quality video productions, you'll need to record good audio *and* good video.

The microphone (or mike) is the basic audio component in any recording system. Most portable video cameras have a built-in microphone (Fig. 3.11). Like any microphone, the built-in mike changes sound into electronic signals. In a portable video system, the built-in microphone sends the electronic audio signals to the videotape deck. The deck then records the sound signals alongside the video signals on the videotape, where both signals are stored for later playback.

Although the built-in microphone is very easy to use, it may not always record the sound you want to record. Some built-in mikes have an "omni-directional" pick-up pattern. This means that the mike picks up sounds coming from all directions around the microphone head (Fig. 3.12). Because of this, the built-in mike generally works best when you are trying to record random or unfocused sounds (such as crowd noises). The built-in mike doesn't do as well with sounds coming from a single point in front of the camera (such as the on-camera voices of an interview) because the omni-directional pick-up pattern tends to gather in unwanted sounds from other sources. Since the built-in mike is part of the camera, these unwanted audio signals can often include the mechanical sounds you create as you use the lens controls and camera trigger.

This problem is sometimes made worse by the AUTOMATIC GAIN CONTROL (AGC) feature found inside many portable video decks. This

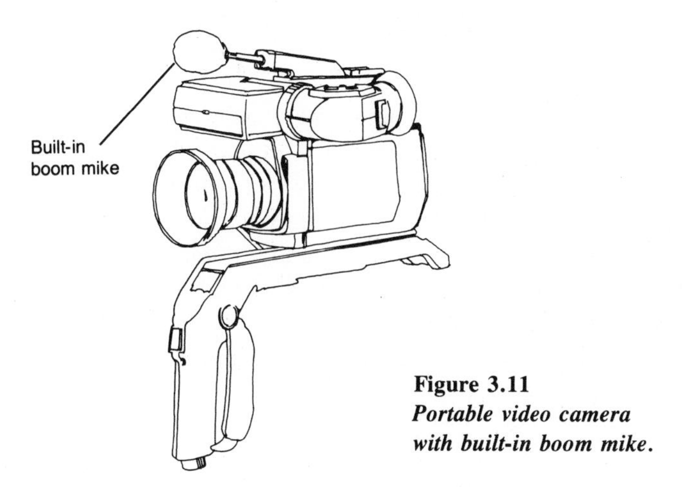

Figure 3.11
Portable video camera with built-in boom mike.

control tries to boost all sounds coming into the camera to a recordable level, and it can come in handy when you are trying to pick up sounds coming from several sources that are located at different distances from the camera. For example, if you were using the built-in mike to record the voices of six people sitting around a table, the AGC would help boost the voices of the people sitting farthest from the camera. Without the AGC, the people farthest from the built-in mike might sound very faint. But the AGC will also boost the level of sounds you may not want on your tape—such as car horns, the sounds of doors opening and closing, and other distracting noises. And when there is a pause in the sound being recorded, the AGC may boost the silence to an annoying recorded hiss.

In Chapter 4 you'll learn how you can use *external microphones* and *line audio sources* to help solve these unwanted audio problems. For now, try to record the best sound possible with your camera's built-in mike. Try to reduce stray or unwanted noises during your video record-

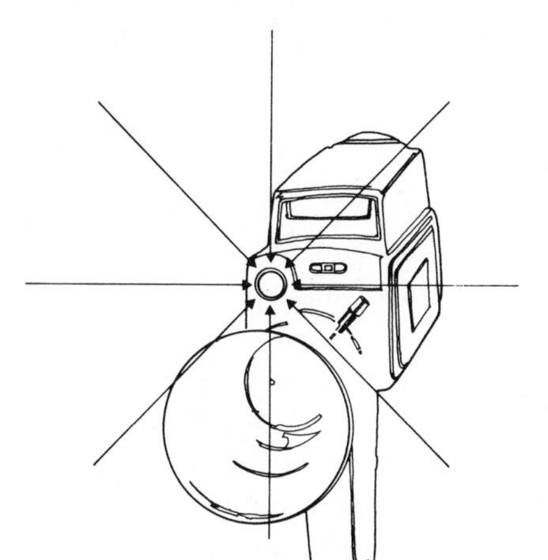

Figure 3.12 ***A built-in microphone with an omnidirectional pickup pattern. Picks up sound from all areas around the mike–even sounds you might not want to record.***

ing sessions, and practice operating your camera quietly to keep the built-in mike from picking up camera sounds.

Some newer portable video cameras feature an AUDIO ZOOM CONTROL on their built-in microphone. Using this control, you can eliminate many unwanted sounds by zooming in on the audio coming from one area within a recording location. Check to see if this control is available on your camera's built-in mike. If it is, check your operator's manual to see how it works, and use it to help eliminate unwanted audio in your video productions.

Keeping It Running: Basic Care of Your Portable VTR

Bumps and Bruises If you're like most portable video producers, you'll shoot many of your video productions outdoors. For each outdoor shooting session, you'll need to pack up your video system and carry it to the recording location. Usually, you'll also need to set up your video system without the tables and chairs that help support the equipment during indoor productions.

The people who manufacture portable video systems know this, so most manufacturers try to make sure that their equipment is rugged enough to handle the minor bumps and bruises that are a normal part of outdoor recording sessions. But that doesn't mean you can neglect or abuse your video system. Like any electronic equipment, your video equipment can be damaged by rough handling, and it needs your help to stay in good running condition.

In caring for your portable video system, your best guide is common sense. No video system likes moisture or temperatures that are too hot or too cold, so keep your equipment away from these conditions. Protect your video equipment from rain and moisture, keep it away from direct sunlight, and keep it warm during cold weather. Here are some other suggestions:

- When you're transporting a video system to a recording location, try to protect the equipment from heavy jolts as much as possible. Hard knocks can damage the circuits inside the deck and loosen screws on the camera, deck, and AC adapter. Never leave any part of your system locked in a car or van, especially in hot or cold weather.

- If you notice loose screws, tighten them carefully with a small screwdriver. Then you can keep them tightened by lightly dabbing nail polish over the tightened screws.
- If you have a carrying case or protective vinyl cover for your video system, use it whenever you transport or store the equipment.
- Be sure to follow any special maintenance instructions listed in your operator's manual.

Always use special care with your camera—the most delicate piece of equipment in any video system. Handle the camera gently, and never leave it where it might be bumped or dropped. Most importantly, be sure to protect the camera's pickup tube from bright bursts of light. You can do this by observing three simple rules.

1. Always replace the lens cap when the camera is not in use.
2. Always turn the aperture ring to the *highest* F/stop setting (or the *C* setting) when the camera is not in use.
3. Never point the camera at the sun or any bright shining object (light bulbs, candle flames, chrome car bumpers on a sunny day, and so on).

Cleaning the Video Heads When you play back videotapes, you'll sometimes see streaks of static (or "snow") in the image displayed on the TV screen. This problem is usually caused by dirt or metal oxide that has built up on the video heads (Fig. 3.13). If only one head is dirty, half the picture may be snowy. If both heads are dirty, the whole image may be snowy.

There are two ways to clean the video heads on a videocassette recorder. The first and hardest way is to remove the top casing of your deck and clean the heads with special swabs and head-cleaning fluid. A

Figure 3.13 ***If streaks of static (or "snow") appear on the TV screen as you play back a tape, it's probably time to clean your deck's video heads.***

second, easier way is to use a *head-cleaning cassette* that cleans the heads by running a specially coated tape through the deck.

Because the first method requires removal of the deck's casing cover, we don't recommend it. The inside area of a video deck is unknown territory to most of us. Unless you're a trained video technician, you risk damaging the deck when you remove the casing and rummage around inside. You may also void your manufacturer's warranty.

Try the second method instead. Buy or borrow a head-cleaning cassette for your Beta or VHS deck and use it when you see snow on your TV screen as you play back a tape. Be sure to follow the directions on the head-cleaning cassette, *and don't use the cassette any more than you have to*. The tape on most head-cleaning cassettes is coated with a rough substance something like sandpaper. Using a headcleaning cassette too often can wear down the video heads.

Once a year you should bring your video system into a repair shop or equipment store for a complete cleaning and checkup. During the checkup a technician will clean the tape transport path, adjust and lubricate the deck's moving parts, and clean the video, audio, and erase heads. The technician will also use a special demagnetizer to rid the deck of any stray magnetism that may have built up on the audio and video heads.

Problems?

Like all video producers, you'll occasionally run into problems during video recording or playback. Sometimes the problem is caused by a damaged component or a broken videotape. More often, the problem is caused by a minor, easily corrected mistake.

If you run into a problem, first check to see that the equipment is getting power and that each component connection is correct and secure. You should also double-check to see that you've set the function and camera controls correctly. If this double-checking doesn't solve the problem, try reading over the trouble-shooting suggestions in your operator's manual.

If none of these steps helps, check over the trouble-shooting guide on the following pages.

Table 3.1 Trouble-Shooting Guide

Problem	*Possible Cause*	*Possible Cure*
No power		
With internal battery	Internal battery disconnected	Connect battery properly
	Internal battery discharged	Recharge battery
	Internal battery dead	Replace battery
	Internal battery connectors damaged	Inspect and repair (if possible)
With external battery	DC (4-pin) plug, cable, or jack damaged	Inspect and repair if possible
	DC plug not properly connected	Inspect and connect
	External battery discharged	Recharge battery
	External battery dead	Replace battery or choose another power source (AC adapter or car cigarette lighter adapter)
With AC adapter	AC wall plug not supplying power	Use another AC outlet
	AC cord not plugged in	Plug in AC cord
	AC adapter power switch is off	Turn AC adapter power switch on
	DC plug cable or jack damaged	Inspect and replace or repair if possible
	AC adapter not connected to VTR	Connect AC adapter to VTR
With car battery cord	Improper polarity of car battery cord (indicator on adapter card is lit)	Reverse polarity of plug as shown in operating manual (if possible)
	DC plug or jack damaged	Inspect and repair if possible
	Car battery cord not connected to VTR	Connect car battery cord to VTR
	Cord not properly connected to car cigarette lighter socket	Inspect and connect properly
Loss of power	Causes listed under **No power**	Cures listed under **No power**
	Low battery charge	Check battery meter and replace or recharge battery if necessary
	Disconnected power cable	Check all power cable connections and reconnect as necessary

Table 3.1 *(Cont'd.)*

Problem	*Possible Cause*	*Possible Cure*
No tape movement in RECORD mode (You've pressed the **RECORD** and **PLAY** buttons, but the tape won't move.)	Camera not connected to VTR	Connect camera to VTR
	Camera trigger not on	Pull and release camera trigger
	Camera trigger not working (handle may be loose)	Inspect and correct if possible
	PAUSE or **STILL** control on deck jammed	Inspect and correct if possible
	Multi-pin camera plug, cable, or jack damaged	Inspect and repair if possible
	Moisture (condensation) on tape track (caused by VTR going from cold to warm area)—check for **DEW** light	Allow VTR to warm up and dry out (one to two hours)
No tape movement in PLAYBACK, FAST FORWARD, or REWIND modes	**PAUSE** or **STILL** control jammed	Inspect and repair if possible
	Moisture (condensation) on tape track (caused by VTR going from cold to warm environment)—check for **DEW** light	Allow VTR to warm up and dry out
Electronic camera viewfinder remains dark when recording	VTR hasn't had time to warm up	Allow VTR and camera 20–30 seconds to warm up
	Multi-pin plug, cable, or jack damaged	Inspect and repair if possible
	Camera not connected to VTR	Connect camera to VTR
Electronic viewfinder lights up but fails to show image (or you have no image in a through-the-lens viewfinder) **when recording**	Lens cap not removed	Remove lens cap
	Aperture closed	Open aperture
	Camera picture is of uniform surface or out of focus	Focus on an object at wide-angle setting
	Not enough light	Provide more light

Table 3.1 (*Cont'd.*)

Problem	*Possible Cause*	*Possible Cure*
Electronic viewfinder lights but fails to show image when playing back a tape	Video heads dirty	Clean video heads
	No image recorded	See **No image recorded** section directly below
No image recorded	During recording, tape was not moving because camera trigger was off	Rerecord, making sure camera trigger is on and tape is moving
	Wrong tape is being played back	Make sure all tapes and boxes are correctly labeled and select right tape
	Tape accidentally erased	Rerecord, label tape, and keep recorded tapes away from magnetic fields (electric motors, airport metal detectors, and so forth)
No image on TV screen when playing back a videotape	TV set or deck not turned on	Turn TV and deck on
	Deck not connected to TV	Connect RF line from deck to TV
	BRIGHTNESS control on TV turned too low	Turn up **BRIGHTNESS** control
	TV tuned to wrong channel	Make sure deck and TV set are tuned to the same channel (usually Channel 3 or 4)
Poor image on TV during tape playback (noises, snow, roll, jitters, dropout)	Tracking problem	Adjust tracking control (slowly)
	Video heads dirty	Clean video heads
	Video heads magnetized	Have technician demagnetize heads
	Low battery (if you are using battery power for deck)	Recharge battery or use alternative power source
	Interference from broadcast television signals	Change deck and TV to channel that is not used by a station in your area. Disconnect TV's internal or rod antenna

Table 3.1 (*Cont'd.*)

Problem	*Possible Cause*	*Possible Cure*
Dark spots or lines on image viewfinder (spots or lines remain stationary on viewfinder)	Dirt on viewfinder	Clean viewfinder with cleaning swab
	Dirt on eyepiece lens	Clean eyepiece lens with lens tissue
	Dirt on front of camera pickup tube	Have technician clean front of tube
	Pickup tube is burned (burns are areas of the tube desensitized by bright concentrated light)	Point camera at a smooth white surface, defocus, set proper aperture, and record for up to half hour (this procedure may reduce or remove minor burns, but it also shortens tube life)
Poor color during playback of videotape	TV set color controls need adjusting	Adjust TV's **COLOR, HUE,** and **PICTURE** controls until color looks natural
	Imbalanced color recorded on videotape	Adjust camera's **WHITE BALANCE** and **COLOR TEMPERATURE** controls and rerecord program
No sound or weak sound recorded	Sound on TV not turned up or working properly	Turn up TV volume control or use another TV set if necessary
	Dirty audio heads	Have technician clean audio heads and rerecord
	Audio heads magnetized	Have technician demagnetize audio heads and rerecord
	External microphone not properly connected to VTR	Properly connect external microphone to VTR and rerecord
	Battery in external microphone (if so powered) is weak or dead	Replace battery and rerecord
	Multi-pin plug, cable, or jack damaged (carrying sound from built-in camera mike to VTR)	Inspect and repair if possible

When you go out shooting, wear comfortable, loose-fitting clothes and good shoes. Nonslip rubber-soled shoes or work boots are good. While recording or playing back, be careful not to put any strain on connecting cables, plugs, or jacks. Disconnect your components before transporting them, even from place to place during a shooting session. Whenever possible, do your shooting with at least one other person. One-person shooting is possible, but it's not convenient, and it can lead to costly mistakes. When shooting on location, use a camera extension cable when you can so that if you have to move around to get the best shots, you only have to move the camera.

Beat the Zoom (A game to improve your zooming skills)

The zoom ring is one of the most useful features on your video camera. Earlier in this chapter you learned how to use the zoom ring to zoom in and zoom out on a scene. In Chapter 4, you'll learn how you can use the zoom ring to set up and "frame" the different shots that make up a video production. This game will help you understand how the zoom ring works and help you to learn to use the zoom ring smoothly and effectively.

Beat the Zoom is an outdoor activity for two or three people. Here are the materials and equipment you'll need:

- One complete portable video system for recording the tape (a camera, deck, and fully charged battery)
- A deck, TV set, and power supply for playing back the tape
- A blank videocassette
- A camera tripod (not absolutely necessary, but it will help)

How You Do It This game uses nothing more than a simple single-camera portable VTR. It doesn't matter whether the portable video system is black and white or color. If there are three people, one of you can operate the camera, one can operate the VTR deck, and the other can perform as the camera subject. If there are only two, one person can be responsible for both the camera and the deck. The activity can be repeated once or twice so that each member of the team has a chance to do the required jobs.

In the game, the camera operator and the camera subject have to work together. The idea is for the camera operator to use the zoom ring on the camera lens to keep the subject's size from changing as he moves toward or away from the camera.

The camera should be set up on a tripod if possible. This will keep the camera steady throughout the game. The portable video system should be connected to its power source and the power should be turned on. Put the system in the RECORD/STANDBY mode, making sure that you get a clear image in the viewfinder. Point the camera away from the sun at a long open area, such as an athletic field or a long straight path. Set the lens aperture at the correct number (if you are using the manual aperture control) and adjust the zoom lens for its tightest (or maximum) telephoto setting.

Once the camera is set up, the camera subject walks into the camera's viewing range. With direction from the camera operator, the subject places himself far enough away so that his feet rest exactly at the bottom edge of the viewfinder frame, and so the top of his head is placed precisely at the top edge of the frame. It will take a couple of minutes for the camera subject to find the correct position, but with a little patience and communication, this won't be difficult.

When the camera operator tells the subject that he is in the right position, the camera operator (or his helper) should put the VTR deck in the RECORD mode. When the camera operator says "forward," the camera subject begins walking very slowly straight toward the camera. At the same time, the camera operator turns the zoom ring to the right so that the subject's feet remain exactly at the bottom edge of the viewfinder frame and the head remains at the top of the frame. In other words, the camera operator should keep the size of the camera subject from changing within the viewfinder frame as the subject walks toward the camera. The only way to do this is by turning the zoom ring. As camera operator, you may also have to tilt the camera up or down slightly. When your subject gets close enough so that you can't turn your zoom ring anymore, ask him to stop.

Once you've had some practice, you can quicken the pace and make the activity a little more competitive. To start, the camera operator and subject should cooperate, because you'll need some practice. Later on, though, the camera subject can move faster and reverse direction without warning from time to time, to see if the camera operator can keep up with him. Obviously, the subject shouldn't move from side to side. If he does, the whole point of the exercise is lost.

What to Look For There are several good observations to make when you play back the tape you've recorded. First, if you were the camera operator, you can see how successful you were at keeping the subject's body size from changing. Or if you were the subject, you might note how often you moved out of the camera viewfinder's rectangle.

What else should you look for? Take a look at the subject as the zoom ring was turned from its tightest setting to its widest angle. Do the subject's features look flatter at one end of the zoom range than at the other? Do the shadows on the subject seem to change? How about the picture's foreground and background? Where in the zoom range do the foreground (the area in front of the subject) and the background (the area in back of the subject) seem to be in sharp or soft focus? What effects do a sharp or soft background focus have on the main subject? When would you like your main subject's background and foreground to be in sharp or soft focus? Why?

chapter 4

MAKING IT BETTER: SKILLS AND TECHNIQUES TO IMPROVE YOUR VIDEO PRODUCTIONS

Shots, Scenes, and Video Transitions

If you have all the components of a portable video system, you have all the tools you need to build your own video productions. But you'll also need some building materials. That's where shots, scenes, and video transitions come in. Shots and scenes are the pieces that make up a video production, just as words, sentences, and paragraphs are the building materials that make up a written message.

In this chapter, you'll learn what the different camera shots are and how to decide which shots will work best for the video message you're trying to convey. You'll also learn how video *transitions* can help you move from shot to shot, how *extra lights* can help brighten up your shots, and how simple *special effects* can help you get your message across. Finally, you'll learn how *audio dubs* and *external microphones* work and how you can piece together (or edit) shots and scenes to build a finished video program.

Camera Shots

There are four basic shots in video production: the long shot, medium shot, close-up, and extreme close-up. As you plan your video productions, it's important to remember that each of the basic shots serves a different purpose. The following list describes each of these shots.

- A long shot (LS) shows your subject from a long distance.
- A medium shot (MS) brings the viewer in closer, so much of the background that is present in the long shot disappears.
- A close-up (CU) shot brings your viewers very near your main subject.
- The extreme close-up (XCU) shows one subject, or one part of a subject, in very close detail.

Figure 4.1 A LONG SHOT (LS) SHOWS YOUR SUBJECT FROM A LONG DISTANCE. *To set up a long shot, you zoom out to a wide-angle setting on your zoom lens. The long shot is sometimes called an* **establishing shot,** *because it helps establish (or set) a scene and background for viewers. For this reason, video producers often use a long shot at the beginning of a production.*

Figure 4.2 A MEDIUM SHOT (MS) BRINGS THE VIEWER IN CLOSER, SO MUCH OF THE BACKGROUND THAT IS PRESENT IN THE LONG SHOT DISAPPEARS. *This shot shows subjects without a camera. To set up a medium shot, zoom to a middle (or medium) setting on your zoom lens. Since a medium shot includes less background to distract viewers, the subjects (not the background) become the main focus for viewers. The MS usually shows people from the hips or waist up and is usually "tight" (or close) enough so viewers can recognize people's faces.*

Figure 4.3 A CLOSE-UP (CU) SHOT BRINGS YOUR VIEWERS VERY NEAR YOUR MAIN SUBJECT. *To set your camera for a close-up, zoom in to (or almost to) the maximum telephoto setting on your zoom lens. The close-up will usually show a person from the shoulders or neck up. With a close-up, your main subject seems to fill the screen, so there is very little on the screen to distract viewers' attention from the main subject.*

Figure 4.4 THE EXTREME CLOSE-UP (XCU) SHOWS ONE SUBJECT, OR ONE PART OF A SUBJECT, IN VERY CLOSE DETAIL. *In an extreme close-up, you use your camera almost like a microscope. You bring the camera in extremely close to the subject, and you zoom all the way in to the maximum telephoto setting. Because the XCU shows no background at all, it forces viewers to pay full attention to the details of the person or object that fill the TV screen. For this reason, video producers often use an extreme close-up of a subject's face to show very strong emotional reactions (fear, surprise, anger, etc.). Video producers also use extreme close-ups to focus viewers' attention on important parts of a person or object.*

Setting Up Your Shots

Framing When you frame a shot in video, you arrange the scene and adjust the camera so viewers see exactly what you want them to see. A well-framed shot is one whose borders highlight the main subject in a pleasing way. Here are some simple suggestions to help you frame your shots:

Figure 4.5

LEAVE HEADROOM AT THE TOP OF THE SHOT. ***When you frame a human subject in your viewfinder, leave a small amount of headroom between the top of the subject's head and the top edge of the frame. A good guideline is to place the subject's eyes about two thirds of the way up the frame. This will usually leave enough, but not too much, headroom.***

Figure 4.6

IN PROFILE SHOTS, PLACE YOUR SUBJECT OFF TO THE LEFT OR RIGHT OF THE FRAME. ***If the profile shot shows the subject looking to the right, place the subject off to the left side of the frame. If the profile shot shows the subject looking to the left, place the subject off to the right side of the frame.***

Figure 4.7

AVOID PLACING YOUR SUBJECT IN THE EXACT HORIZONTAL CENTER OF THE FRAME. *Usually, a shot will look better if you frame the main subject slightly to the left or right of center, rather than exactly in the center of the frame.*

ARRANGE GROUPS OF SUBJECTS IN DIAGONAL OR TRIANGULAR PATTERNS, RATHER THAN IN STRAIGHT LINES. *Arranging subjects in diagonal lines or triangles adds depth to your shots. Because a TV screen tends to exaggerate the amount of space between people, try to group your subjects closer together than they would be in a real-life situation.*

Figure 4.8

Figure 4.9

USE NATURAL FRAMES AS MUCH AS YOU CAN. *When you shoot outdoors, tree trunks and other natural frames can make your shots seem more interesting and attractive. Place your natural frames on the left, right, top, bottom, or any combination of borders in your viewfinder.*

Figure 4.10A NORMAL (STRAIGHT) ANGLE *In a straight-angle shot, set the camera roughly at eye-level directly in front of your subjects. The subjects will usually face the camera directly, as if they were speaking to your viewers rather than to each other.*

Playing the Angles When you frame a shot, you will also have to decide on a camera angle. If you decide on a "normal" camera angle, you'll set the camera directly in front of your subjects and show viewers what they would see if they were standing right in front of the scene. If you choose a different camera angle, the shot will have a very different effect on your viewers. The photographs on these two pages show four basic camera angles and describe the effect each angle has on viewers.

Figure 4.10B SIDE (PROFILE) ANGLE *In a side-angle shot, you set the camera around eye-level and off to the left or right side of your subjects. The profile angle usually shows the subjects speaking to each other and makes your viewers feel that they are looking in on the scene without being noticed by the subjects.*

Figure 4.10C HIGH-ANGLE SHOT *In a high-angle shot, you set the camera above your subjects and shoot down on the scene. The high-angle shot makes your subjects seem smaller and less powerful than they are in real life. Because of this, the high-angle shot gives viewers a feeling of power and command over the scene.*

Figure 4.10D LOW-ANGLE SHOT *In a low-angle shot, you place the camera below your subjects and shoot up at the scene. The low-angle shot makes your subjects seem larger than they are in real life. Because of this, the low-angle shot is very useful when you want subjects to seem powerful and important to your viewers.*

Getting From One Shot to the Next: Transitions and Editing

Once you've decided on the shots you'll use in a video production, you must think about how you'll move from one shot to the next. That's where *transitions* come in. In video, a transition is a way of moving smoothly from one shot or scene to the next. Transitions are the "glue" that connects different shots and scenes into a complete edited production. The following pages describe some useful video transitions and suggest ways you might use each transition in editing your portable video productions.

Generally, the best transitions are the ones your viewers never notice. If your transitions go unnoticed, you've probably been careful to keep your camera movements smooth and steady, and you've probably done a good job of working your transitions into the flow of your production.

Zoom Transition The zoom transition is an easy way of moving from one basic shot to another when there's no break in the action. For example, many productions begin with a long, or "establishing," shot that sets the scene, followed immediately by a medium shot that shows people and objects in more detail. Using your camera's zoom lens, you can make this transition by zooming in from the long shot to the medium shot. Of course, the zoom transition works just as well with the zoom moving the other way. For example, if a scene calls for the camera to move from a medium shot to a long shot, you could just as easily use the zoom ring to complete this "zoom out" transition.

For best results, your zoom transitions should be very slow and steady, and they should be used sparingly. A zoom that is too fast or rough is distracting and annoying to your viewers. Many professional producers never use a zoom during a shot. They only use the zoom lens to set up their shots.

Stop-and-Start Transition The stop-and-start transition is one of the simplest ways of moving from one scene to the next. Because you'll be stopping and starting the tape, this transition works best when there is a clear break in the action between shots or scenes. The stop-and-start transition is sometimes called a *straight cut*.

Here's how it works. As the action in the first scene comes to an end, pause the tape by pulling the camera trigger. With the tape in the PAUSE (or RECORD/STANDBY) mode, quickly set up the shot that will begin your next scene. When everything is ready, pull the camera trigger again. With the second pull of the camera trigger, the tape should start moving and the deck should be recording again.

The stop-and-start transition is easy to do, but it does have some drawbacks. First, because it takes a moment for the videotape to get up to the right recording speed once it starts moving again, pulling the trigger the second time can cause a small "glitch" when the first scene ends and the second scene begins. Also, since keeping the deck in PAUSE will drain your battery and wear down the surface of your videotape, you'll have to move to the next scene in a hurry. If you'll need more than two or three minutes to set up the next scene, you should switch the deck from PAUSE to STOP. Unfortunately, this can cause even more of a glitch when you start the tape recording again.

There's a fairly easy way of avoiding this extra glitch. First, if you know you'll have to stop between scenes, record a few seconds beyond the end of the first scene before you push STOP. Just before you're ready to start recording the next scene, put the tape in REWIND for a few seconds and replay the end of the first scene. When you reach the point where you want the scene to end, put the deck in PAUSE and press the RECORD button on the deck again. (On some decks, you may have to press both the PLAY and RECORD buttons at the same time to activate the RECORD function.) Make sure the RECORD, PLAY, and PAUSE lights are all lit. If they are, you're ready to record again with a pull of the camera trigger, and you'll avoid the extra glitch you sometimes see when you start recording from the STOP position.

Even though you're starting from PAUSE, you may still see a small glitch between scenes. For smoother transitions and for transitions that suggest that time has passed between the scenes, try the black-to-black, defocus/refocus, and cardboard-wipe transitions described here:

Black-to-Black (or Fade) Transition The black-to-black transition works like the stop-and-start transition, but it puts a short stretch of black between scenes. As the action in the first scene ends, slowly turn the camera's aperture ring to the highest F/stop setting (usually marked

F/16 or *C*). As you turn the ring, the image being recorded will slowly grow dimmer. When you've turned the aperture ring to the *C* or highest F/stop setting, the image should be faded to an even black. At this point, pull the camera trigger to put the deck in PAUSE and get ready to record your next scene.

Just before you're ready to begin recording the next scene, focus the scene in your camera viewfinder (with the deck and camera in RECORD/STANDBY), and return the aperture ring to the highest (or closed) highest F/stop setting, the image should be faded to an even black. At this point, pull the camera trigger to put the deck in PAUSE and get ready to record your next scene.

When you play the video recording back, you'll see that the first scene ends with a slow "fade down" to black, and that the next scene begins in black and slowly "fades up" to a bright and balanced image. Because the pausing and starting of the tape took place while the image was faded to black, you'll barely notice the small glitch that comes between the scenes. Because it places a few moments of black between scenes, the black-to-black transition suggests that time has passed between the two scenes.

The black-to-black transition is only possible on a camera that has an aperture ring or automatic fade control. If your camera has an aperture ring *and* an automatic iris control, you'll need to use the automatic iris override switch (see page 47). Some newer video cameras have an "automatic fade" feature that fades to black by pushing a button.

Defocus/Refocus Transition This transition is like the black-to-black transition with one important difference. Instead of turning the aperture ring to fade the image up and down, you turn the lens focus ring to defocus and refocus the image. As the action in the first scene ends, you zoom all the way in and slowly turn the focus ring until the image dissolves to an unfocused blur. Then you pull the camera trigger to pause the recording process and with the camera and deck in the RECORD/STANDBY mode, you go ahead and set up your next scene.

Just before you're ready to begin recording the next scene, focus the scene in your camera viewfinder and set the aperture ring at the correct F/stop setting (if your camera has an aperture ring). Then turn the focus ring until the image is an unfocused blur and pull the camera trigger. With the deck recording again, turn the focus ring until the image is in sharp focus.

When you play back the video recording, you'll see that the first scene slowly dissolves to an unfocused blur. After a few moments of the blurred image, the next scene will gradually come into focus. Because the tape stops and starts when the image is a blur, the glitch between scenes won't be very noticeable. Like the black-to-black transition, the defocus/refocus transition suggests that time has passed between scenes.

Cardboard-Wipe Transition With the cardboard-wipe transition, you end the first scene by sliding a piece of black or gray cardboard in front of the lens. When the cardboard completely covers the lens (so you see only black in your viewfinder), pull the camera trigger to put the deck in PAUSE. Then with your camera in RECORD/STANDBY, set up your next scene, making sure that it looks well balanced and focused in your viewfinder. When you're ready to begin recording, slide the cardboard back in front of the lens (so you see only black in the camera viewfinder), and pull the camera trigger to start the recording process again. Once you're recording, slide the cardboard away from the front of the lens. With the cardboard removed, you should see a shot that's bright and well focused.

The cardboard-wipe transition works best when you slide the cardboard from left to right or top to bottom with a steady hand and keep the cardboard as close to the front of the lens as possible.

When you replay a cardboard-wipe transition, you'll see that the first scene seems to be pushed (or wiped) off the screen as the cardboard moves from top to bottom or left to right. Then, the black you see for a moment seems to be wiped off the screen by the image that begins the second scene. Like the black-to-black and defocus/refocus transitions, the cardboard-wipe transition is most effective when there is a clear break in the action between scenes and you want to suggest that time has passed from one scene to the next.

Keeping the Camera Steady

Sometimes it's very hard to keep a video camera steady during a recording session. For example, if you are videotaping the action in a fast-breaking news story, you might not have time to steady or brace yourself for smooth camera work. Later, when you play the tape back, the picture may seem to shake and sway on the TV screen.

Occasionally, these unsteady images can add an interesting "newsreel" feel to your portable video productions. Most of the time, though, an unsteady picture will simply distract and annoy your viewers. To avoid this added distraction, try to keep your camera steady and stable during video recording sessions. Your productions will look better, and your viewers won't get seasick watching the screen!

Getting Help From a Tripod Although there are many different kinds of video camera tripods, they are all designed to do the same thing: to take the weight of the camera off your hands and put that weight on the three legs that form the tripod's base. Since a tripod's legs don't twitch or grow tired (as your own hands, arms, and legs sometimes do), a tripod can be a great help when you're trying to record steady images with your video camera. In fact, we recommend that you buy or borrow a tripod and add it as a basic accessory to your portable video system.

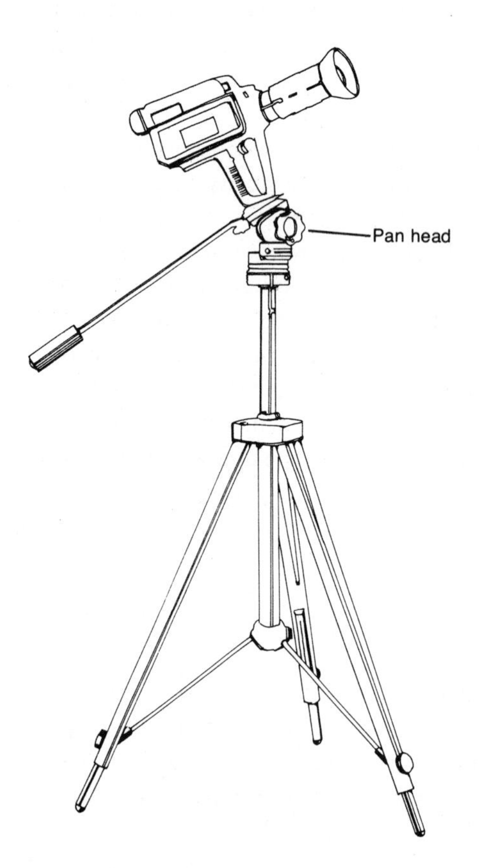

Figure 4.11
Portable video camera attached to tripod with "pan head."

There are a few things you should be sure to check when you shop for a video camera tripod. First, be sure to check the tripod's *weight rating*. Some tripods were designed to hold movie or still-photo cameras, and they may not be strong enough to support the weight of your video camera. You should also check to make sure the tripod has a *pan head*. If the tripod doesn't have a pan head, you won't be able to tilt and pan the camera during a recording session (see pp. 82 and 83).

Attaching your camera to the tripod is easy. With most cameras, you simply screw the bolt on the top of the tripod into the threaded hole you'll find at the bottom (or base) of your camera. On some video cameras, this tripod "jack" will be at the bottom of the camera's *pistol grip* (Fig. 4.11).

Keeping a Steady Hand Although a tripod is handy to have, there will always be times when you want or need to hold the camera by hand. Here are some ways you can help keep your shots steady during hand-held camera work.

ANCHOR YOUR FEET AND GRASP THE CAMERA'S PISTOL GRIP WITH YOUR WRITING HAND. ***If you are shooting from a standing position, think of yourself as a human tripod. Like any tripod, your main support comes from your legs. Try placing your feet firmly on the ground, about a shoulder width apart, with one foot slightly in front of the other. This should give you a firm, comfortable foundation for steady camera work. If you're right-handed, you should grasp the pistol grip with your right hand. If you're left-handed, you should grasp the grip with your left hand. When you position your hand on the grip, make sure you can reach the camera trigger easily with your index finger.***

Figure 4.12

Figure 4.13

KEEP ONE ELBOW TUCKED IN AND MOVE FROM YOUR WAIST AND HIPS. ***As you operate the camera, keep the right elbow (if you're right-handed) or the left elbow (if you're left-handed) tucked in to your body. This will help keep the camera steady. When you need to move or pivot the camera, keep your arms and elbows still and move from your hips and waist.***

WHENEVER POSSIBLE, BRACE YOUR BACK AGAINST A SOLID OBJECT. ***In most production situations, you can get extra support by bracing your back against a wall, a tree, or some other stable object.***

Figure 4.14

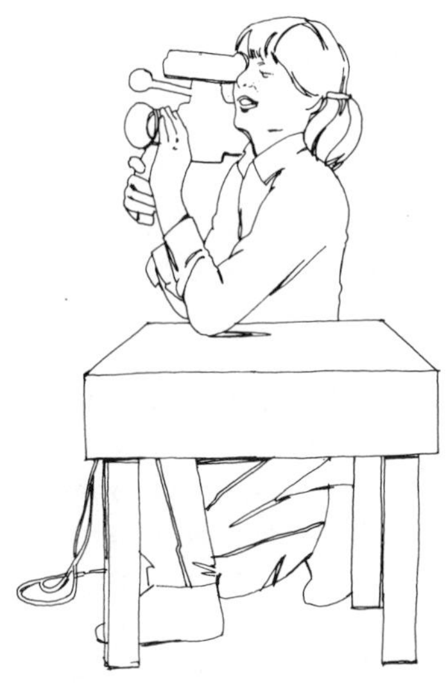

Figure 4.15

WHENEVER POSSIBLE, BRACE YOUR ELBOW ON A FLAT SURFACE. ***If you rest the elbow of the arm that's holding your camera on a table, the hood of a car, or some other flat surface, you'll be able to hold your camera more steadily than when you're simply standing up. You can also get good camera stability by lying on your stomach and resting your elbow on the ground or floor.***

WHEN SHOOTING FROM A CROUCHED POSITION, REST YOUR ELBOW ON YOUR KNEE. ***This will help steady the camera and keep you from falling over.***

Figure 4.16

Sooner or later, you'll find yourself in situations that require you to experiment with other ways of supporting your video camera. For example, if you were shooting a program about a small child, you might want to lie flat on the floor and shoot up at the scene, to show the world the way the child would see it. Or you may want to sit on the floor and shoot the scene with your elbows resting on a coffee table. As long as you have time to set up a shot, you should be able to find a way to support and steady the camera.

More Camera Movements

You already know about one camera movement: the *zoom*. An explanation of how to use your zoom lens is on page 49. There are four other camera movements you may find useful in portable video production: the *dolly, truck, tilt,* and *pan*.

Dollying Like the zoom, the dolly moves you closer or farther from your subject. When you zoom, you move in or out on the subject by adjusting the zoom ring on your camera lens. When you dolly, you move in or out by moving the entire camera in (dolly in) or out (dolly out).

In TV studios, cameras are usually mounted on wheeled tripods (called dollies). With a wheeled tripod, you dolly by simply rolling the camera toward or away from your subjects. With a portable video camera, smooth dollying isn't so easy. Since portable video cameras are not normally mounted on wheeled tripods, you dolly a portable camera in or out by walking the camera toward or away from your subjects. As you walk, your leg and body movements will make it hard to keep the camera steady.

Trucking When you truck, you move yourself and the camera to the left (truck left) or to the right (truck right). Usually, you truck the camera to get a different angle on a scene or to follow your subjects as they move to the left or right.

Tilting In the tilt, the camera operator swivels the camera up (tilt up) or down (tilt down). Like most camera movements, tilting is more of a body movement than a camera movement. When you tilt the camera, you should brace your elbow (the elbow of the arm that is holding the camera's pistol grip) in to your body and move the camera up and down by moving your waist and neck. Tilts are useful for following up-and-down action and for establishing high-angle and low-angle shots. (Figure 4.16 shows a camera operator "tilting up" to frame a low-angle shot.)

Panning When you pan the camera, you pivot the camera to the left (pan left) or right (pan right). For a smooth and steady pan, you should hold the camera still and slowly pivot to the left or right by moving your hips and waist. Throughout the pan, you should keep your elbow tucked in to your waist and your feet firmly planted in the same spot. Pans are useful for following action as it moves to the left or right or for gradually revealing subjects that are arranged in lines or rows.

Getting Tricky: Simple Special Effects for Portable Video

In video, a *special effect* is a way of presenting an image in a unique or unusual way. Professional TV producers use special effects all the time. For example, when you watch a TV show on one of the commercial networks, you may see one image placed, or superimposed, over a second image. Or the producer may split the screen in half to show you a different image on each side of the screen. Professional TV producers create these effects with an electronic component called a *special effects generator*. Although they can create many interesting images, special effects generators are too expensive to be used by most portable video producers.

But you don't need expensive electronic machinery. You can create the following simple special effects with the components of your portable video system and with ordinary supplies found in many schools and homes.

Figure 4.17 ISOLATION SHOT ***The isolation shot pinpoints (isolates) one object or person within the scene. To create an isolation shot, place a cone of rolled-up black construction paper in front of the camera lens. This will black out the rest of the scene and focus viewers' attention on the subject in the center.***

Figure 4.18 SOFT EDGE *The soft edge highlights one object or person within the scene by blurring the edges of the image. To do this, use a clear plastic lens filter and some petroleum jelly. Apply the petroleum jelly to the edges of the filter, leaving a small hole in the center. Then, attach the filter to the front of the lens.* **Be careful not to get petroleum jelly on the lens itself.** *Later, clean the lens filter by wiping the petroleum jelly with a tissue soaked in turpentine or paint thinner and then rinsing the filter with mild detergent and warm water. You can also try covering the lens filter with waxed paper instead of petroleum jelly. Cut a hole in the center of the waxed paper and tape it to the lens filter.*

Figure 4.19 SOFT SCENE *In a soft scene, the whole image is slightly blurred. To create this effect, cover your lens with a piece of cotton gauze, an old nylon stocking, a piece of window screening, or a piece of cheesecloth. Experiment with different materials until you create a soft feel that looks good to you.*

Video and photographic supply stores often sell special-effects lenses as add-ons to video cameras. The two special-effects lenses most often used in video are the close-up and the multiple-image lenses.

A word of warning. When you buy or borrow a special-effects lens, check to make sure that the lens will fit onto the front of your camera's lens. If it doesn't, a video or camera supply store may have adapters that will adjust the lens for the proper fit.

Although you shouldn't be afraid to experiment with different special effects, you should always remember one important rule: Don't use a special effect unless it serves a clear purpose in your program. If you use special effects just for the sake of using them, you'll just be adding unnecessary clutter to your productions. This can confuse your viewers and distract their attention from the message you're trying to get across.

Figure 4.20 A CLOSE-UP LENS *acts as a magnifying glass, allowing you to videotape very small objects or small parts of objects. When you use a close-up lens, use a tripod to keep the magnified image steady.* A MULTIPLE-IMAGE LENS *multiplies a camera's subject so it appears in several places on the screen. Usually, the label on a multiple-image lens will tell you how many times it will multiply the subject.*

Sounding Better: Ideas to Improve Your Audio Track

Although your portable video camera's built-in mike is very convenient, it may not always record the sound you want. As explained in Chapter 3, most built-in mikes have a very wide pickup pattern. Because of this, a built-in mike tends to pick up all the sounds occurring in or around a recording location, whether you want to record the sounds or not. In this section, you'll learn how *external microphones* can help you record only the sounds you want on your tape. You'll also learn how to perform *audio dubs* and how you can use *audio line sources* to add special sounds and effects to your video productions.

External Microphones An external microphone offers several advantages over your portable camera's built-in mike. First, because there are many different types of microphones to choose from, you can select the type that best fits your recording situation. Second, since you are able to use the external mike with a microphone extension cable, you'll also be able to move the microphone in close to your subjects. With the mike in close, the subjects' voices will sound strong and clear, and you shouldn't pick up many of the unwanted sounds the built-in microphone picks up when the camera is located more than a few feet from the sound source. (Plugging in an external microphone cuts off the camera's internal mike.)

It's up to you to decide whether you want the extra convenience that comes with using the video camera's built-in mike, or the extra sound quality that comes with using an external microphone. If you decide to use an external microphone, here are some suggestions for recording the best possible sound.

- *Buy or borrow a good quality general purpose microphone*. We suggest that you try a good quality *dynamic*-type microphone. (*Condenser*-type mikes record slightly better sound, but they are more expensive and require batteries.)
- *Choose a mike with the right pickup pattern for your needs*. For general use, a dynamic mike with a *cardioid pickup pattern* usually works best. A cardioid mike picks up sounds in a heart-shaped pattern around the microphone. An *omnidirectional pattern* picks up sounds equally from all directions surrounding the

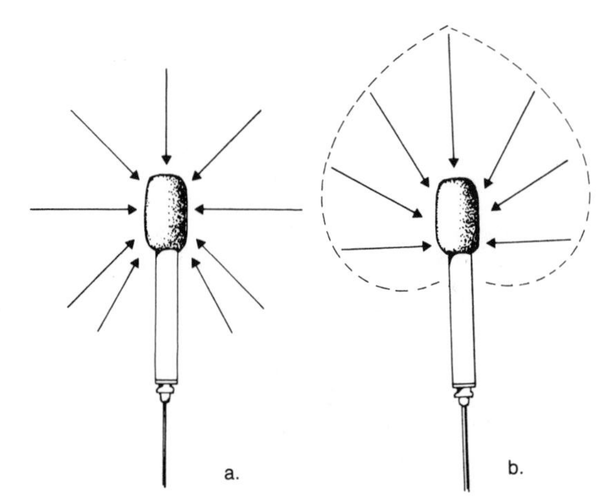

Figure 4.21
A microphone with an omnidirectional pickup pattern picks up sounds from all areas around the mike. A microphone with a cardioid pickup pattern picks up sounds from a heart-shaped area in front of the mike.

mike. A unidirectional pattern picks up sounds only from directly in front of the mike.

- *Make sure that your extension microphone is a low impedance mike*. This mike will match the impedance of the microphone input on your portable video recorder. (The impedance rating should be written on the microphone. A low impedance rating ranges from 300 to 600 ohms.)
- *Make sure that the plug on the end of the microphone cord will fit into the microphone or mic in jack on your deck*. If the plug doesn't fit, you'll need to buy an audio plug adapter at an audio or video supply store.
- *Use a mike extension cord when possible*. For many recording situations, a microphone extension cable and mike table stand will come in handy.

Figure 4.22
Crew member monitoring audio during a video production session.

- *During a video recording session, monitor the sound through a headset or earphone*. All portable video decks have a headphone or earphone jack. By plugging headphones or an earphone into this jack, you can make sure that the external mike is working and that you're recording only the sounds you want to record.

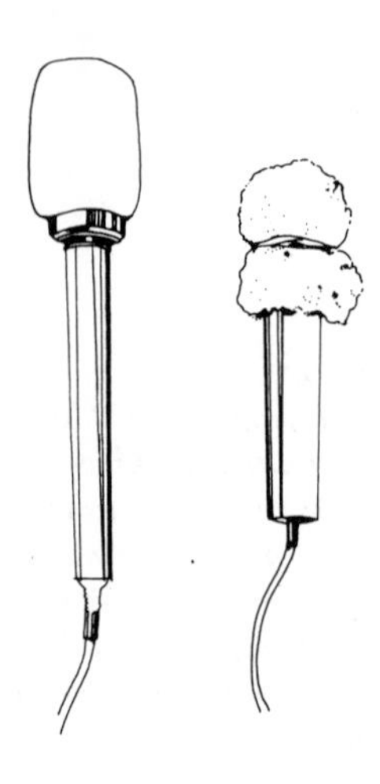

Figure 4.23
Microphones with "store-bought" and homemade windscreens.

- *Use a windscreen.* This is a piece of rubber or plastic foam that wraps around the head of a microphone. Windscreens help filter out the "pops" you sometimes hear when a subject speaks too close to the mike, and they help eliminate the rumbling sound sometimes caused by wind at a recording location. You can buy windscreens at video or audio supply stores, or you can make your own with foam or cotton and a rubber band. (A windscreen is sometimes called a "pop filter.")

Figure 4.24 ***Microphone held between interviewer and subject.***

- *In an interview, hold the microphone between the subjects, about 6–12 inches below mouth level.* With the microphone placed as illustrated, you should hear both subjects loud and clear, and you won't need to wag the mike back and forth between them.

Figure 4.25
Performer with microphone placed correctly for good sound.

- *Have performers speak over, rather than directly into, the microphone*. This too will help prevent the "pops" you sometimes hear when performers speak too close to the microphone.

Figure 4.26 ***Performers placed in a semicircle, with microphone in the middle. (Note that microphone is placed on a table stand, on top of a foam pad.)***

- *Arrange large groups of people in a semi-circle, and place the microphone in the middle*. This will help to pick up individual voices and keep members of the group from sounding too far away.
- *When you use a microphone table stand, place the stand on a pad of sponge, foam, or paper*. This will help keep the mike from picking up the sound of people tapping their fingers or pencils on the table, as shown in Figure 4.26.

Audio Dubs You can use the audio dub control on your portable video recorder to add music to a videotape, narration to shots recorded at a remote location, or to replace parts of an audio track that didn't turn out quite right. When you use the audio dub, you'll be erasing and rerecording only the *audio* part of the tape. The video part will stay the same.

A good audio dub takes some planning. First, you'll have to decide what sort of audio you want to add to the tape. In one type of audio dub, you plug a microphone into the MIC IN jack on your deck and add the new audio track by speaking into the mike. In a more complicated kind of audio dub, you add the new sound by plugging a record player, an audio tape deck, or some other *audio line source* into your video deck. Using *patch cords*, you connect the tape recorder or record player to the AUDIO IN or LINE INPUT jack on the video deck. Once you've made this connection, you can dub music, sound effects, or anything else recorded on an audio tape or record.

Be sure to connect the tape deck or record player into the AUDIO LINE-IN or AUDIO-IN jacks on your deck and *not* to the microphone input. Audio line sources send out a much stronger signal than microphones, and sending such a strong signal into the microphone input can damage your deck's audio circuits. If you're not familiar with audio or stereo equipment, you may need some help connecting the audio tape deck or record player to your portable video recorder.

For an easier way of dubbing sound from an audio tape or record, try plugging a microphone into your video deck and placing the microphone in front of a speaker that is playing back the sound from the audio tape deck or record player. This method may add a small amount of noise to your audio dub, but you won't have to spend time hunting for patch cords and figuring out how to connect the audio line source to your VTR.

Once you've connected a microphone or audio line source to your deck, performing an audio dub is just a matter of pushing the right buttons. Here are a few simple steps to follow:

1. *Connect the components.*
2. *Find the right spot on the videotape.* Play back your prerecorded videotape until you reach the point where you want the audio dub to begin. To find the right spot, watch the tape on a TV screen or your camera's electronic viewfinder. When the tape reaches the right point, press your video deck's PAUSE control.

3. *Get ready to record the new sound.* If you're using a microphone, turn down the volume control on the TV set to avoid *audio feedback*. If you're using an audio line source, make sure the record or tape is set (or cued) at the point where the new audio begins.
4. *Press the audio dub control on your VTR.* Once you do this, both the AUDIO DUB and PAUSE controls should be switched on.
5. *Perform the audio dub.* Press and release the PAUSE button (so the tape is moving once again) and begin your audio dub. If you're using a microphone, begin speaking into the mike at this point. If you're using an audio line source, begin playing the audio tape or the record. If you wish, you can *fade* up your recorded sound at the beginning of the dub and back down again at the end. When you reach the point where you want the audio dub to end, press the STOP control on your VTR.

Finally, you should be sure to rewind the tape and check your audio dub to make sure everything turned out all right. If you notice problems, go back to Step 1 and do the dub over again.

Basic Lighting for Video

In the earliest days of video, lighting was often a major problem. The earliest color video cameras required a great deal of light, and small changes in lighting conditions would disrupt the camera's complicated color balance system. Today, most portable video cameras can operate with much less light, and most feature color balance controls that are quick and easy to use. With these changes, portable video producers can now spend less time worrying about how much light falls on a scene and more time thinking about ways they can use light to create the right mood and the best possible image.

Outdoor Lighting During the daytime, you should have no trouble finding enough outdoor light for your portable video productions. Even on cloudy days, there should be more than enough sunlight to create a clear, bright image in your camera viewfinder. In fact, cloudy days often provide the best lighting conditions for portable video production. When the sun is too bright, you may find too much contrast between the bright and dark areas in your shots. Clouds help prevent this problem by diffusing sunlight and softening shadows.

Here are a few suggestions that can help put your outdoor shots in the best possible light.

- Set your camera's color temperature dial or lens filter control at the correct setting (daytime, evening, cloudy daylight, etc.).
- Use your camera's aperture ring (if it has one) to soften harsh contrasts. If objects look too bright and shadows look too dark, adjusting the aperture ring to a higher F/stop number can help bring the image into better balance.
- Avoid shooting with the sun behind your subjects. They will appear very dim compared to the bright light behind them. Some portable video cameras have a back light switch that helps correct this problem.
- Avoid pointing the camera at direct or reflected sunlight. As you know, bright flashes of light can damage the camera's pickup tube.
- If possible, arrange the scene so your subjects appear separated from their background. You can do this in two ways: by placing the subject in front of a contrasting background (a girl in a dark blue dress against a light yellow background, for example), or by lighting the subject more brightly than the rest of the scene. In outdoor productions, try moving the subject into the brightest areas of the scene. You can also try using cardboard covered with aluminum foil to reflect sunlight onto your performers.

Indoor Lighting Many of the suggestions listed under Outdoor Lighting also apply to indoor lighting. When you shoot indoors, it is still important to make sure that the camera's color temperature control is set at the right position, that you have avoided placing your subject directly in front of a bright light, and that you have arranged your subjects so they appear separated from the background. Of course with indoor productions, you'll be missing one important element found in outdoor recording sessions—the sun.

When you shoot indoors, you'll usually need to replace the sun's natural light with artificial light. Professional TV producers use very expensive artificial lights that imitate the brightness and color quality of sunlight. Unless you can borrow a professional light kit, you'll probably have to make do with whatever room lights you find at your indoor location or whatever inexpensive lights you've been able to bring along.

When you arrive at an indoor shooting location, you should check to see whether the light available from ceiling lights, table lamps, and windows is bright enough to produce a clear video image with good color quality. If it is, you may not need extra lights. If the available light is too dim, or if it produces poor color quality, you'll need some sort of extra illumination. Be careful, though. Always remember that extra lights consume extra electricity. Be sure that you don't overload electric circuits by plugging too many lights into too few wall sockets.

Here are two low-cost (and safe) suggestions for using artificial lights in your indoor video productions.

- *Buy or borrow inexpensive clip-on lights*. They have a spring-action clamp that allows you to attach them to door frames, window ledges, and other convenient locations. Clip-on lights use an ordinary light bulb. You can buy them in many department, photographic, and video supply stores. We suggest that you buy or borrow at least one as a basic accessory to your portable video system. For many indoor locations, one clip-on light may be all you need to brighten up your shots!
- *Experiment with the three-point lighting arrangement*. In this system, one light spotlights the front of the subject, a second light helps fill in the shadows created by the spotlight, and a third light illuminates the top and back of the subject's head and shoulders. Professional TV crews use three different types of

Figure 4.27
An inexpensive clip-on light.

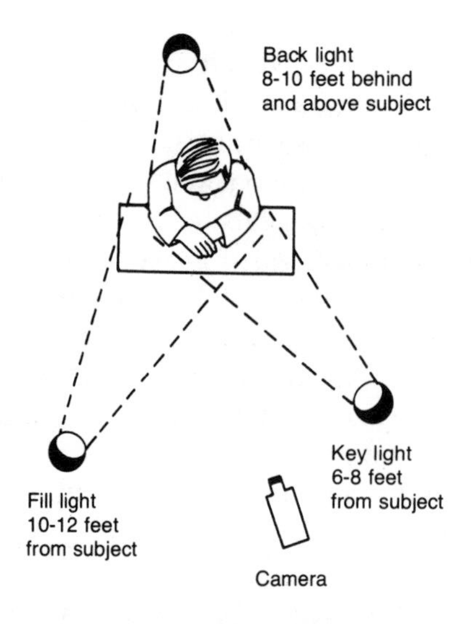

Figure 4.28
Three-point lighting arrangement.

lights in this arrangement, but you can try it with three clip-on lights. Place one above and to the right of your subject, so it is 6 to 8 feet from the subject and shines on the subject's face at a slight angle. This will be your main, or *key*, light. Place a second one above and to the left of the subject and a bit farther away (10 to 12 feet) than the first light. This *fill light* should fill in some, but not all, of the shadow created by the key light. Place the third light behind and 8 to 10 feet above the subject's shoulders. This *back light* will highlight the top and back of the subject's shoulders and help separate the subject from background scenery. Check this lighting arrangement in your camera's viewfinder and record a short test segment. If the image doesn't look quite right, rearrange the lights until it does.

Always avoid shooting with a window directly behind your subject. Having a bright window behind your subject is like having the sun at the back of the scene. Like the sun, the window will show up fine, but your subject may seem too dark. The back light switch found on some video cameras can help correct this problem, but you can usually avoid the problem altogether with a little planning. Try to arrive at your recording location early and arrange furniture and people so your subjects won't be standing in front of windows once the recording session begins. And again, always record a short stretch of test tape to check your lighting quality before recording.

chapter 5

PLANNING AND PLOTTING FOR VIDEO

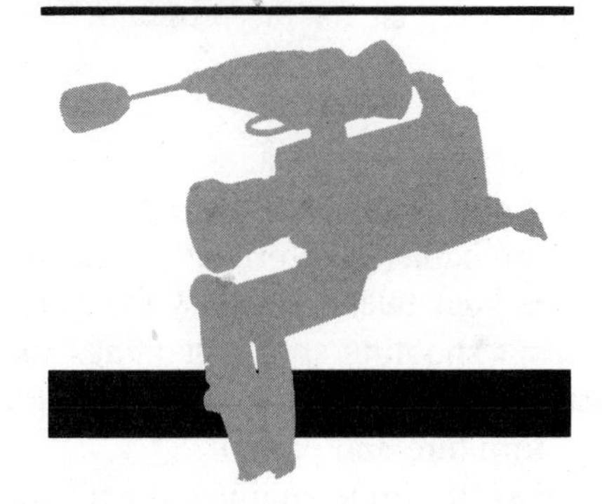

Ways to Organize Your Video Planning

Careful planning is the best way to guarantee a high-quality video production. In video, planning is called pre-production. Professional video producers spend lots of time on pre-production, often more time than they spend on the production session itself. Usually, time spent on pre-production is more than paid back in time saved on shooting and in the higher quality of your programs.

During pre-production, you organize every aspect of the upcoming production. First, you determine why you want to produce a program in the first place. Then, you think through the content of your shots and determine where each shot will appear and how long each will last. You'll also need to write a script for any spoken parts and to complete any background research for the program. Here's a list of other pre-production tasks:

- select shooting locations
- find or build props and scenery
- choose on-camera performers and a shooting crew
- schedule your talent and crew for shooting sessions
- check out shooting sites (for things like location of AC wall outlets, placement of windows for lighting, and availability of useful furniture and props)
- make lists of all the equipment and supplies you'll need
- plan and prepare any graphics or artwork you'll need

In video, you usually work as part of a production team, so pre-production tasks are shared among members of the team. Obviously, some kinds of video production require more pre-production planning than others, but every production session will need some planning. Every production session also needs a *producer*—the person who leads the production team and divides the pre-production tasks among team members.

Professional video producers use one or more methods to organize their shooting sessions. Several of these methods are explained on the following pages. Look them over and adapt them for organizing your own portable video work.

Table 5.1 Producer's Checklist I

Your name (producer) ______________________
Name of program ______________________
Description of shooting session ______________________
____
Date of shooting session ______________________
Location of shooting session ______________________

Equipment needed (mark yes or no for each item as you pack for shooting)
_____ Video camera
_____ VTR portable deck
_____ TV set for playback
_____ Charged batteries (when were they charged?) ______________
_____ AC adapter/battery charger
_____ External microphone
_____ Earphone or head phones (for monitoring sound as you record)

Supplies needed (mark yes or no for each item as you pack for shooting)
_____ Blank cassettes (how many?) ________
_____ Video head-cleaning cassette
_____ Scripts and other production plans
_____ Clipboard
_____ Paper and pencils
_____ Duct tape (for securing microphone and camera cables)

Cables and accessories needed (mark yes or no for each item as you pack for shooting)
_____ Camera tripod or camera brace (specify which) ______________
_____ Extension microphone cable
_____ Extension camera cable
_____ AC extension cord
_____ Spare battery
_____ RF connector cable
_____ Car battery adapter cord
_____ Tool kit
_____ Other (specify) ______________________

Condition of equipment (mark yes or no before you leave for shooting)
_____ Have you connected and tested every piece of equipment you'll need to make sure it is working properly?

Table 5.2 Producer's Checklist II

Performers' names	*Performers' roles*
______________	______________
______________	______________
______________	______________
______________	______________

Production crew (write the names of the persons who do the production jobs listed)

______________ Director

______________ Camera operator(s)

______________ VTR operator(s)

______________ Sound director

Special supplies (list all needed supplies and check as you pack for shooting)

Graphics/titles	Check
______________	____
______________	____
______________	____
______________	____
______________	____
______________	____
______________	____
______________	____

Props/sets/furnishings	
______________	____
______________	____
______________	____
______________	____
______________	____
______________	____
______________	____
______________	____

Music/other sound sources	
______________	____
______________	____
______________	____
______________	____
______________	____

The Producer's Checklist To start your planning, you should have a work sheet for listing all the people and things you'll need for your shooting session. You might call this a producer's checklist. In Table 5.1, we've provided an example of a checklist you could use to organize portable production sessions—both at home and on the road. The second page of the producer's checklist is more concerned with the people and the materials needed to support your production.

A producer's checklist helps you to organize materials and people for a shooting session. It does *not* help you organize what happens once you get there. For this extra organization, you'll need to try one of the three basic methods for writing and arranging a video shooting session:

1. shooting scenario
2. script
3. storyboard

Before trying any of these three methods, you should know a little about the scripting abbreviations and production hand signals that experienced video producers use in writing and shooting from a scenario, script, or storyboard.

Hand Signals and Abbreviations

Hand signals and abbreviations make it much easier for directors, scriptwriters, and producers to communicate with their performers and production crews. During a production session, a director (the person in charge of the shooting session) or camera operator usually uses hand signals to send messages to on-camera performers. An interviewer, for example, has to know when to begin talking and when time is running out. If the director of a shooting session spoke these instructions out loud, his voice would be picked up by the microphone. With hand signals, the director is able to give instructions quickly *and* quietly.

Hand Signals Professionals use dozens of hand signals during the course of a shooting session. For most portable video sessions, the following seven key signals should be enough.

1. Cue Talent (begin performing now)
   With outstretched arm point finger emphatically at performer or performers.
2. Speak Up
   Lift up the palms of both hands repeatedly. (Indicates that the performer's voice is not being picked up.)
3. Speak More Softly
   Push palms of both hands downward repeatedly. (This is the direct opposite of the speak-up signal.)
4. Look at the Camera
   Point to the camera with an exaggerated over-the-head arm motion. (Signifies that the performer has looked away from the camera.)
5. Stretch It Out
   Pull fingers apart repeatedly from the center of chest, as though pulling taffy. (Indicates that there is still lots of time left which should be filled with ad-libbed remarks.)
6. Wrap It Up
   Roll one arm over the other repeatedly, as if you were calling a traveling foul in basketball. (Indicates that time is drawing to a close and that there are only 15 seconds to the end of the shot.)
7. Cut
   Draw index finger emphatically across your throat. (Signifies that time has run out and performer should stop right away.)

Figure 5.1a ***Cue talent!***

Figure 5.1b *Speak up!*

Figure 5.1c *Speak more softly.*

Figure 5.1d *Look at the camera!*

Figure 5.1e *Stretch it out!*

Figure 5.1f *Wrap it up!*

Figure 5.1g *Cut!*

Abbreviations Scriptwriters use abbreviations to indicate commonly used instructions. These abbreviations are also useful with shooting scenarios and storyboards—they save you the trouble of fully writing out every instruction each time it is used. Here are some abbreviations that will be helpful for portable video production:

C = Camera
LS = Long shot
MS = Medium shot
CU = Close-up
XCU = Extreme close-up
PL = Pan camera left
PR = Pan camera right
MIC = Microphone
TL = Truck camera left
TR = Truck camera right
DI = Dolly camera in
DO = Dolly camera out
TU = Tilt camera up
TD = Tilt camera down
ZI = Zoom camera lens in
ZO = Zoom camera lens out

Take C = Start camera
Fade ↑ = Fade up—open the camera image up from black
Fade ↓ = Fade out—close the camera image down to black
Roll Tape = Start the VTR recording
Standby = Get ready to begin the session

Scenarios, Scripts, and Storyboards

The Shooting Scenario Shooting scenarios are simpler and less formal than scripts or storyboards. Scenarios are very good for organizing on-location documentaries and video interviews. They are also useful when precise timing and a spoken script are not required. The shooting scenario is a rough outline that tells you where and how to set up your camera, where your on-camera performers should be, how your props and furnishings should be arranged, and how your shots should be sequenced (put in order).

Usually, you write a shooting scenario after a pre-production visit to your shooting location. A pre-production visit allows you to note important characteristics of the location—such as the location of windows and AC power outlets, the pattern of shadows at outdoor locations, the arrangement of furniture, and the dimensions of the rooms where you plan to do your shooting. You should note all of this information on your shooting scenario.

You can write scenarios on ordinary 8½″ x 11″ loose-leaf paper or scratch pads. You use them *with*, not instead of, the producer's checklist described earlier. Very often, scenarios are also used with scripts, which are more formal and give more detailed directions. When used with a script, the scenario would provide the general organizational details for shooting (the basic outline) and the script would give the exact minute-by-minute shooting details.

Let's take a look at a sample shooting scenario. It contains important general information in its heading and is organized into four columns that contain more detailed information about the shooting activity. Our sample scenario describes two segments of an interview at the home of a local beer can collector.

Table 5.3 Shooting Scenario

Production title: ''Metal to Money: Collecting Beer Cans''
Producer: Sabrina Cooper
Location: Living room of Penfield Warren
33 Revere Road, Chelmsford
Date: Thursday, June 12
Time: 3:30 P.M.—after school
Estimated length of program: 10 minutes

Shot number	*Content of shot*	*Time of shot*	*Notes*
1	Shot of three note cards in sequence	4 seconds for first card	There is a large bay window in Mr. Warren's living room facing south, letting in lots of sunlight
	Run 60 seconds of black *leader*. Use manual aperture control	3 seconds for second card	Place the camera on a tripod in the corner of the room to the right of the window
	Open from black slowly on first program title card	4 seconds for third card	There is an AC outlet on each wall, so use house electricity instead of a battery
	Close down to black again on third title card; hold black for 10 seconds		

Table 5.3 Shooting Scenario (*Cont.*)

Shot number	*Content of shot*	*Time of shot*	*Notes*
	Card 1: ''Metal to Money, Collecting Beer Cans'' Card 2: ''Produced by Sabrina Cooper'' Card 3: ''Directed by Matt Goldston''		Place the three title cards in a ringbinder on a stand about six feet away from the camera. Don't let the sun shine directly on the title cards. Place VTR on the end of a table beside the couch in the south-west corner of the room. Camera extension cable might be needed. Also need an AC extension cord at least 30 feet long.
2	George Anderson interviews Mr. Warren Rewind tape to point where shot 1 goes black Start shot from black; keep black between this shot and last shot less than 3 seconds Start with CU of George, then zoom back to get MS of George and Mr. Warren From time to time zoom in on Mr. Warren and back out to MS	About 3 minutes for this part of interview	Remove camera from the tripod. Move it to the other side of the window so that it faces the couch. Camera can be held by hand from a crouching position George Anderson and Mr. Warren sit close to one another on the couch for the interview Set up list of possible interview questions on newsprint pad and easel to the left of the camera

Table 5.3 (*Cont'd.*)

Shot number	*Content of shot*	*Time of shot*	*Notes*
	Begin interview —George introduces self —then introduces Mr. Warren Asks questions like: 1. How many beer cans have you collected? 2. Do you collect all beer cans or only special ones? 3. What is the rarest beer can in the world? 4. What is the rarest beer can in your collection? 5. How long have you been collecting beer cans? 6. Has beer always been available in cans? 7. How much money is your collection worth? Close shot by fading to black. Hold black for 10 seconds.		George holds the extension MIC between Mr. Warren and himself

This is a partial scenario for just over three minutes of a ten-minute program. The rest of the scenario might take the crew (and the viewer) to Mr. Warren's beer can collection, to some outdoor beer can collecting activity at the town dump and along roadways, or to a local beer can swapping fair. You might end by returning to your home base, where you would explain and videotape illustrations from a library book on beer can collecting.

The "Metal to Money" shooting scenario seems to outline a lot of detail. But there is a lot of detail that this scenario has left out. It does not include exact camera shots or the exact duration of the interview. It suggests interview questions, but doesn't dictate exactly what they must be. In other words, this scenario gives the producer, the performers, the camera operator, and members of the crew some room to make spot decisions as they shoot.

The Script Unlike rough shooting scenarios, a script gives specific second-by-second details on every aspect of a shooting session. It tells the performers exactly what to say and when. It tells the camera operators precisely how to set up and shoot. It also tells the VTR operator when to start and stop the video recorder. Finally, the script lists all props, sets, and furnishings and tells where each should be located on the set.

The script coordinates the performance, technical details, and timing in a very strict order. Because of this, they work best for skits, dramatizations, and other tightly organized studio-type productions. Scripts don't work as well for interviews, documentaries, or other less formal productions, where the crew and talent need more freedom to follow the action as it unfolds. In portable video, there may be many times when you will need to write a detailed script for your production, but you won't need to script all of your productions.

A simple scripting form has three columns: one for all the technical commands, one for the second-by-second shooting time, and one for audio. The audio column shows all spoken parts, music, and other sound effects. The technical commands column includes everything else: instructions for camera setup, camera shots, and microphone placement, along with VTR operating instructions and technical instructions for the performers. (As you know, the performers will receive their instructions through hand signals.) On the next two pages we show part of a script for a five-minute spoof on a space/horror television show. It uses many of the abbreviations explained on page 103.

Table 5.4 Script

Production title: "The Amazing Adventure of Dr. Omega"
Producer: David Lawton
Location: School Media Center
Date: October 22
Time: 10:30–11:30 A.M.

Technical Commands	*Time*	*Audio*
Sequence 1		
Use tripod and dolly C CU on title stand		
Black on line		
MIC at phono speaker		
Standby video and audio		
Roll tape	0:00	
Take C		
Fade ↑ music	0:05	Music fades up for both titles and fades back down at end of title sequence
Fade ↑ C on title stand	0:05	
Flip title card	0:08	
Fade ↓ C to black	0:12	
Fade ↓ music and out	0:12	
Stop tape	0:20	
REW to 0:15	0:15	
Sequence 2		
Arrange mock chemistry lab on long 6-foot table in studio		
C LS Dr. Omega working in lab from low angle. Use tripod and dolly		
MIC on lab table hidden from camera by lab apparatus. Dr. Omega makes lots of clinking lab-type sounds		
Narrator hides under table as close to MIC as possible		
Black on line		
Standby video and audio		
Roll tape	0:15	
Take C		
Fade ↑ C on LS lab	0:18	

Table 5.4 (*Cont'd.*)

Technical Commands	*Time*	*Audio*
Cue narrator	0:21	*Narrator* (speaks very slowly): We are now in the laboratory of the amazing Dr. Omega!
Cue Dr. Omega	0:27	*Dr. Omega* (speaks even more slowly):
Slowly DI and ZI to XCU of Dr. O's face	0:28	At last . . . the *ultimate* power! Now I can return to my own home planet, Omega, which by a weird fluke is named after me! (He laughs a crazy laugh.) First, I must find another assistant . . . but he has to be young, for although I'm only
C slow DO and ZO to LS again	0:45	275 years old, going on 276, I'm not as strong as I was 110 years ago. Of course I have Egory, and he's young, but he's also stupid.
	0:53	Egory! Egory! come here!
Egory enters screen right, hunched over and looking grotesque.	0:55	*Egory* (breathlessly): Yes, master?
	1:02	*Omega* (in a bossy voice): Egory, run and get me the space travel potion!
	1:07	*Egory*: Yes, master!
Egory rushes out screen right, re-enters screen left with glass vial. Dr. Omega still stares after Egory to right.	1:09	
	1:12	*Egory* (breathlessly): Here, master!
Dr. Omega jumps in fright and turns around with a start.		
Dr. Omega grabs the vial from Egory and gulps down the liquid.	1:15	*Dr. Omega* (with evil grin): And now I travel light years
C fade ↓ slowly to black	1:17	through space and time! Heh, heh, heh.
Stop tape	1:25	
REW to 1:20	1:20	

There were five more shooting sequences for this script. Since the shooting was done with a single-camera portable video recorder (an old black-and-white machine), each sequence had to be shot in its proper order. The program started at 0:00 in the time column. The two scripted sequences shown contain one minute and twenty seconds of program time.

Some of the terms in this sample script may seem strange to you. The phrase "Black on line," is a command that tells you to get your portable video system ready to record with a uniform black picture—what you'd have with your camera lens capped or your lens aperture ring set at *C*. "REW" means rewind the tape to the point indicated in the Time column of the script. This is done so that the next sequence can come in quickly after the previous one. For directions on how to move from sequence to sequence in this way, see the section on *transitions* in Chapter 4.

In the "Dr. Omega" script, you can see that every technical and performing detail was organized in a strict time sequence. For every shot, there was a setup instruction. For every voice part, there was an earlier instruction indicating where the person speaking should be and what he or she would be doing. Every single event was timed to the second. (Keeping time is the job of the director or the director's assistant.)

Although a well-scripted video program is a well-organized program, it can also be an over-organized program. If the members of a production crew are too intent on following a script to the letter, a simple mistake can cause the whole session to break down in confusion.

Scripts are great for organizing the details of a production, but they do not provide much help in planning shots for visual appeal. In fact, the earliest television scripts were written for live studio telecasts, when broadcasters couldn't afford to make mistakes and the visual beauty of a studio production wasn't as important as the organizational detail. For a pre-production method that allows you to concentrate more on the visual quality of the program, try using a storyboard.

The Storyboard This is simply an arrangement of cards with rough sketches of camera shots on a shot-by-shot basis. For every new shot you plan, you draw a new card. Once you've drawn a series of cards, you can arrange them on a bulletin board or a large table to get an idea of how your shots flow together. If you don't like what you see, you can rearrange the cards as often as you like. Or you can take out some cards and add new ones.

Film-makers used storyboards long before television even existed. Film-making is generally a one-camera activity. It is impossible to connect two or more movie cameras electronically and switch back and forth among them during a shooting session. Historically, film-makers have been more conscious of the visual artistry of their work than have television producers.

The storyboard is concerned with visual details (or composition) first and organizational details second. With scripts, the opposite is true. Which of the two is better for you will depend on your preference and on the kind of program you are producing. Since video productions communicate at least half of their message visually, it is good to learn to work with storyboards. In any case, the use of a storyboard doesn't prevent you from also using a script. You can use both—the storyboard for planning the visual qualities of your programs and the script for planning the production details.

You can create your own blank storyboard form on standard 8½″ × 11″ ringbinder-size paper that allows you to jot down still more organizational information, as Figure 5.2 shows. With added space for instructions, you can number your shooting scenes in their proper order. You can specify camera angles and note any desired special effects. You can also indicate rough shooting time and describe how you plan to move to the next scene. Even with these added details, though, the storyboard won't let you organize as precisely as you can with a script.

One of the great advantages of storyboarding is that it allows you to arrange and rearrange your scenes right up to production time. Using the fixed form shown in Figure 5.2 might remove this advantage. For this reason, you should probably still do your sketches on index cards or pieces of scratch paper and use them just as you would if you didn't have the form. After arranging your sketches in their proper order, you can then attach them to the middle column of the storyboard form and add your organizational notes (production and audio) afterward.

When you draw sketches for a storyboard, remember that the main idea is to show the basic visual elements of your shot and not to create great pieces of art. Save your artistry (and your time) for your camera work. If your rough sketches fail to provide the detail you think you need

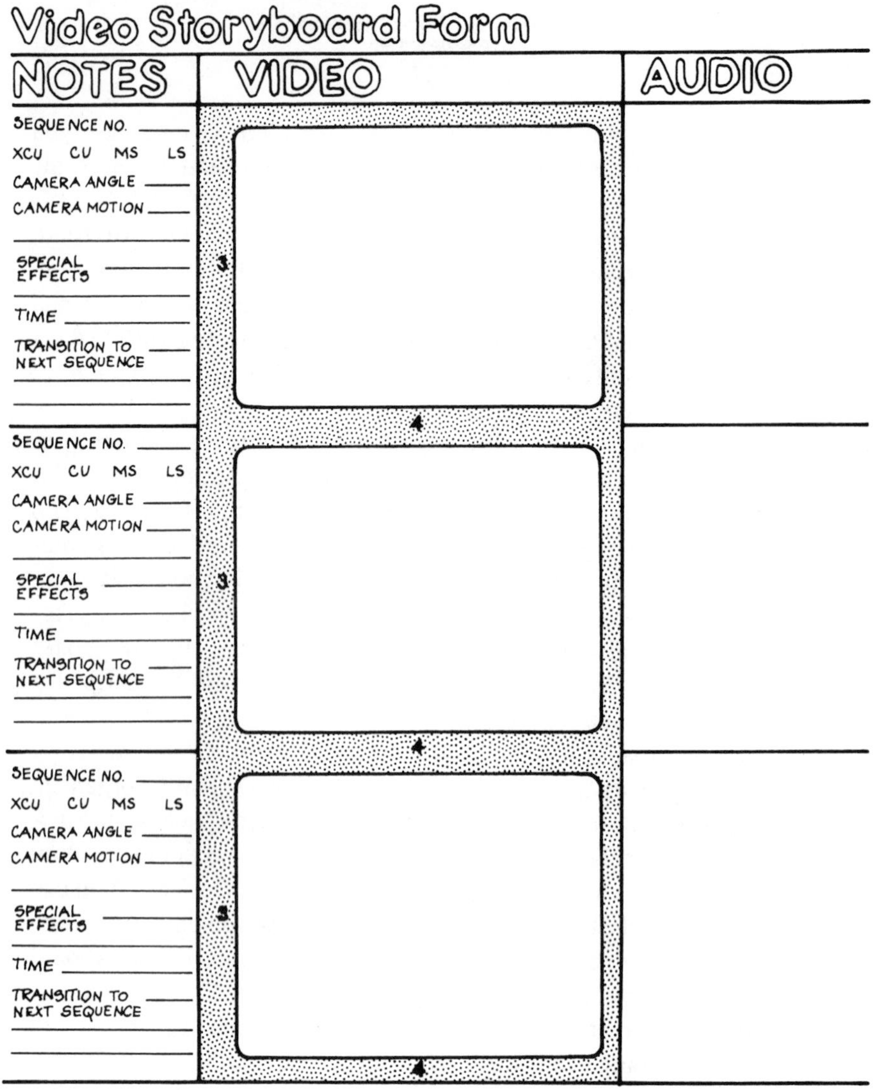

Video Storyboard Form

NOTES	VIDEO	AUDIO
SEQUENCE NO. ____ XCU CU MS LS CAMERA ANGLE ____ CAMERA MOTION ____ SPECIAL EFFECTS ____ TIME ____ TRANSITION TO NEXT SEQUENCE ____		
SEQUENCE NO. ____ XCU CU MS LS CAMERA ANGLE ____ CAMERA MOTION ____ SPECIAL EFFECTS ____ TIME ____ TRANSITION TO NEXT SEQUENCE ____		
SEQUENCE NO. ____ XCU CU MS LS CAMERA ANGLE ____ CAMERA MOTION ____ SPECIAL EFFECTS ____ TIME ____ TRANSITION TO NEXT SEQUENCE ____		

Figure 5.2

in your storyboards, try substituting photographs taken with an instant-type camera for your sketches. If you take the instant-camera approach to storyboarding, however, be ready to spend some money. This type of film is rather expensive, but you can save some money by using black-and-white film instead of color.

Special storyboard pads are available at most art supply and many video stores. They come in various sizes and formats, and they are usually quite expensive. Unless you're very seriously into video production, you can do just as well with your own homemade forms and cards.

Activities to Make You More Comfortable with Pre-Production

Working with shooting scenarios, scripts, and storyboards is not that difficult, but it does take some getting used to. Here are some activities that might help you get started.

Analyze the Elements of Visual Composition As you know, storyboards encourage you to approach your video production visually. But before you start storyboarding, it helps to know what the visual elements of a good picture are. There are certain basic rules which almost all visual artists follow—whether they are video producers, film-makers, photographers, painters, or cartoonists. The way in which visual artists combine the individual elements in a single *image* is called composition. Some of the best examples of good image composition are found in comic strips and magazine advertisements.

Get a slick magazine that you can cut up. (News magazines and publications such as *Life*, *People*, *The New Yorker*, and *Sports Illustrated* are good.) Find strong advertising photographs. Look at them carefully. Try to describe the individual elements as shown in Figure 5.3.

CLINT CLEMENS, BOSTON

Figure 5.3
How has the photographer composed this ad photo for maximum impact

Try also to describe how the photographer's treatment of those elements helps convey a message to you—the person that the advertiser is trying to reach.

To get you started, look at some of the key elements of any photographic image:

- Camera angle (Is the image shown from above or below? From the left or the right?)
- Perspective (Would you describe the image as a long shot, medium shot, close-up, or extreme close-up?)
- Depth-of-field (Is the image's main subject the only thing clear and in focus, or are both the main subject and background in focus?)
- Arrangement (Can you find geometrical shapes or rhythmic lines in the arrangement of objects in the image?)
- Framing (Is the main subject separated from the rest of the picture by tree branches or some other natural frame?)
- Lighting (Does the photographer appear to use lighting effects in some special way?)
- Content (Is the image crowded and busy or is it stark and plain?)

Comic-Strip Storyboards Like advertising photographers, cartoonists are also skilled in using the elements of visual communication in their art work. The color comic strips in Sunday newspapers are very useful for studying the visual elements of images. The "superhero" strips are especially good, because the cartoonists who draw these strips often make exaggerated use of angle, arrangement, and perspective to create a sense of drama and suspense. Next Sunday, try cutting out the individual frames from, say, a *Spiderman* strip. Black out the characters' dialogue with a felt-tipped marker. Then tape the individual comic-strip frames (not necessarily in their original order) onto storyboard sheets, scene by scene (Fig. 5.4).

With your mind now thinking "storyboard" instead of "comic strip," fill in all the "production notes" information blanks on the storyboard form. If your still-image comic frame were to become a shot in a video production, how would the camera move to follow the action? How long would the shot take? What special effects would you use? How would you move from this scene to the next scene? Once you've

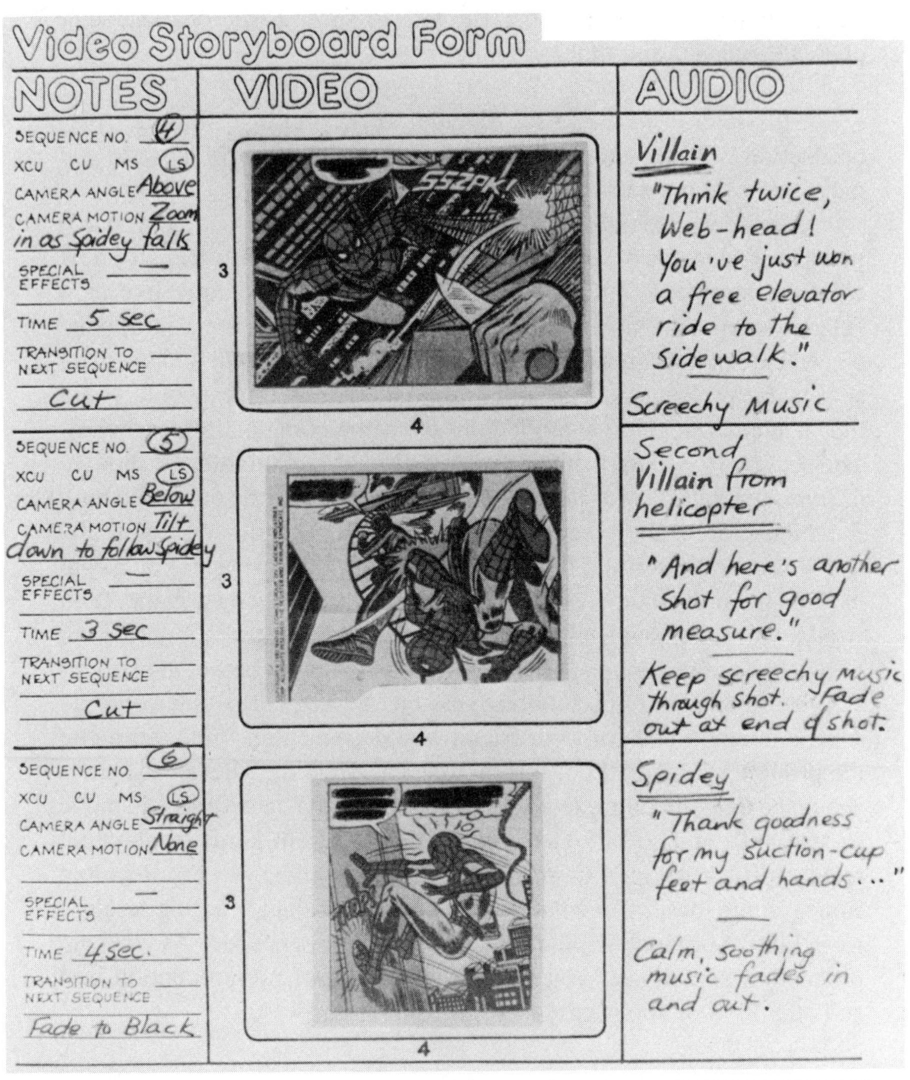

Video Storyboard Form

NOTES	VIDEO	AUDIO
SEQUENCE NO. 4 XCU CU MS (LS) CAMERA ANGLE Above CAMERA MOTION Zoom in as Spidey falls SPECIAL EFFECTS — TIME 5 sec TRANSITION TO NEXT SEQUENCE Cut	3 × 4 frame (SSZPK!)	Villain "Think twice, Web-head! You've just won a free elevator ride to the sidewalk." Screechy Music
SEQUENCE NO. 5 XCU CU MS (LS) CAMERA ANGLE Below CAMERA MOTION Tilt down to follow Spidey SPECIAL EFFECTS — TIME 3 sec TRANSITION TO NEXT SEQUENCE Cut	3 × 4 frame	Second Villain from helicopter "And here's another shot for good measure." Keep screechy music through shot. Fade out at end of shot.
SEQUENCE NO. 6 XCU CU MS (LS) CAMERA ANGLE Straight CAMERA MOTION None SPECIAL EFFECTS — TIME 4 sec. TRANSITION TO NEXT SEQUENCE Fade to Black	3 × 4 frame	Spidey "Thank goodness for my suction-cup feet and hands..." Calm, soothing music fades in and out.

Figure 5.4

answered these questions, try developing a story line, filling in the "audio notes" for each scene. Be sure to show narration, character dialogue, music, and other sound effects.

Storyboards to Scripts As you work your way through pre-production, starting with a storyboard or shooting scenario, how do you move on to organize and write a detailed script?

A good way to find out is to take one of the comic-strip storyboards you just created and develop a detailed script from it. The idea is to show what the story line outlined in the storyboard would look like as a detailed script that includes instructions for camera setup, microphone placement, camera shots and movements, VTR operation, and the parts spoken by performers. As an extra twist, two or three people could independently develop a script from the same comic-strip storyboard. This way, you'll learn from each other. One of you might have thought of some technical details or an extra special effect that the others did not.

There are some types of video production that do not require the use of scenarios, scripts, or storyboards at all. For instance, you might have a friend videotape your tennis stroke while playing a game. You would view the playback to get a better idea of the strengths and weaknesses of your shots. Obviously, you would not bother to script or storyboard this kind of a shooting session. Nevertheless, you *would* do some planning. You and your friend would spend some time setting up the placement and angle of the camera and deciding whether the shots should be long, medium, or close-up, whether your camera should move or stay still, and so on. The point is that different kinds of shooting require different kinds of planning. Some planning is very detailed; some is quite loose. But whatever the purpose of your shooting session, *some* kind of production planning is almost always needed. As you gain production experience, you'll learn what kind of pre-production best suits the kind of shooting you want to do.

chapter 6

GETTING FANCY: TITLES AND GRAPHICS

Sooner or later, you'll probably want to add titles, credits, or other graphic materials to your productions. In video production, a *graphic* is any still visual image created by writing or drawing or by some electronic method. A graphic can be *alphanumeric* (made up of letters and numbers) or *pictorial* (made up of pictures). Graphics can be maps, charts, graphs, diagrams, drawings, or almost any kind of artwork designed to send a message to your viewers (Fig. 6.1). Graphic images can also be created electronically with the use of computers. By using graphics, you can often transform an ordinary video production into an extraordinary one.

In video production, you can create your own graphic materials, or you can use materials found elsewhere, such as in books, magazines, or newspapers. If you plan to use "borrowed" graphics in a production that you will show outside of your own family or classroom (say over a local cable TV access channel), you should first ask permission. A simple letter explaining who you are, what graphic you wish to use, and why you wish to use it will usually accomplish your purpose.

Very often, do-it-yourself graphics will be better suited to the character of your own productions than materials taken from other sources. You can prepare good-looking graphics quite easily and inexpensively. With a little extra effort, you can prepare graphics that look

Figure 6.1 ***Three styles of video graphics.***

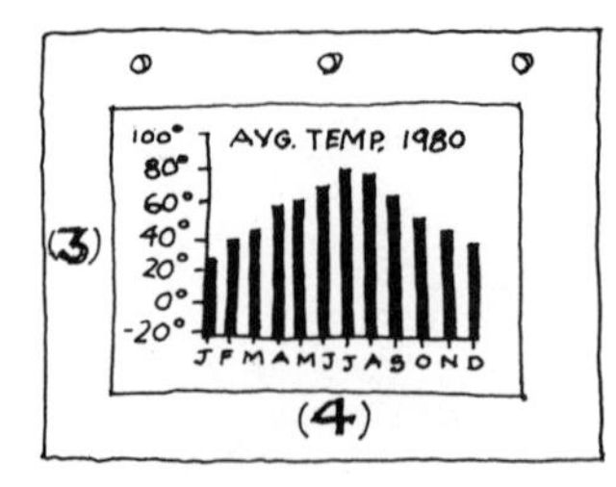

very slick and professional. The rest of this chapter offers suggestions and techniques to help you create graphics for your own portable video productions.

Planning Your Graphics

Timing If you're using your portable video system for a finished production (rather than for viewing and analyzing your tennis stroke, for example), you'll need to plan the order of your shots with some care. You will also need to time them. In more advanced video production, electronic video editing equipment allows you to rearrange segments and add graphics after your original shooting session. Since most of you won't have this luxury, you will have to plan your production to include the necessary graphics in their proper place and in their correct sequence, and then shoot in the order set out in your plan.

As a general rule, you should use a video graphic whenever it would add to a viewer's understanding or enjoyment of your program. For example, if your program is a documentary production about the historical role of American working women, a bar graph somewhere in your program showing the percentage of women in the national work force every five years since the end of World War II might illustrate your point well. Usually, graphics are used at the beginning of a program to show the program title and the major credits (director, producer, important performers). Most films begin this way. Many programs end with graphics, too, showing the production credits in greater detail.

There is no absolute rule about how many graphics should appear, say, in a fifteen-minute program. If the message of your program is made more clear by the use of many maps, charts, and printed information, then by all means use them. You'll learn from experience the best number of graphics to use in any particular production.

How long should an individual graphic remain on a television screen? Again, common sense is your best guide. Graphics with written information should remain on the screen as long as it takes to read them aloud. If this takes longer than five seconds, your graphic may be too crowded and it might be a good idea to break it down into two separate graphics or to create a "graphic scroll" or "crawling" graphic. (We'll explain these later.)

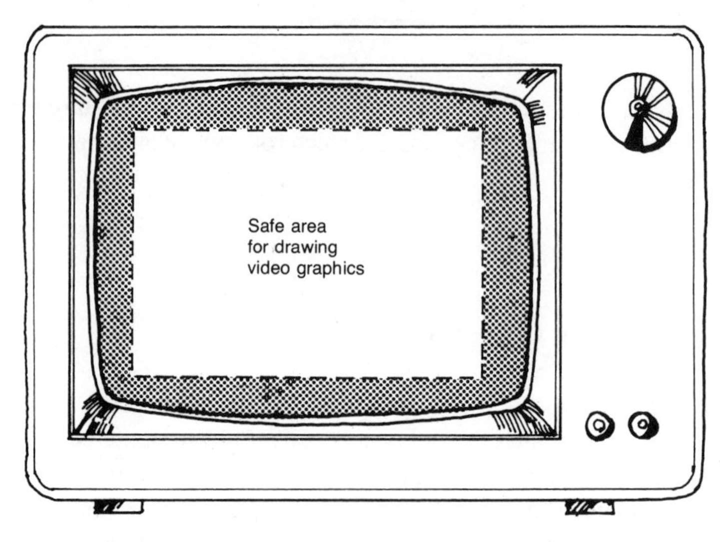

Figure 6.2

Framing Your Graphics When you create graphic materials for video production, you should always remember one important fact: The vertical to horizontal ratio of any television screen is 3:4. That is, for every three inches of height, the screen will be four inches across. In other words, if a TV screen is twelve inches high, it will be sixteen inches wide; if it is nine inches high, it will be twelve inches across. The height/width ratio will always be 3:4, no matter what the absolute size of the screen. (The same is true for the viewfinder of a video camera.)

You should always design your video graphics with this height/width ratio in mind. Otherwise, information at the top or bottom or along the sides of your graphic might be cut out when you try to frame it in your camera viewfinder. The 3:4 ratio of a television screen is called its *aspect ratio*. (Sixteen millimeter movie film also has a 3:4 aspect ratio. On the other hand, the aspect ratio for instamatic cameras is 1:1, a perfect square.) Usually, the absolute size of your graphic material doesn't much matter, as long as it is contained within a 3:4 frame.

There are two other things to remember. Very often, a television screen displays only about 85 percent of the area viewed through a video camera. This 85 percent is called the safe area of a video screen. You should allow for this reduction in size when you create a graphic and whenever you aim your camera at any graphic (Fig. 6.2). By doing this, you'll also be allowing for the slight inward curving of the corners of a television screen. The other thing to remember is to make all graphics the same size if you plan to show them in sequence. If you fail to do this, you will have to adjust your camera's position or its zoom ring every time a new graphic appears for display—a nearly impossible task.

You may find that a 6″ × 8″ (or 15cm × 20cm) rectangle is a convenient size for your graphic work. This size is comfortable for ease of design and artwork, and it's a good size for framing in your camera's viewfinder. A 6″ × 8″ safe area also fits nicely on a standard 8½″ × 11″ sheet of paper or cardboard.

One way to help make sure that all graphic materials fall within a 3:4 aspect ratio is to use a homemade adjustable framing device. Adjustable framers are well known to photographers and film-makers. In their simplest form, they are nothing more than two L-shaped pieces of thick cardboard which, when used together, form rectangles of various sizes (Fig. 6.3). The shorter legs of each *L* represent the vertical dimension of an image, and the longer legs, the horizontal dimension. To make your own framer, start at the inside corner of one of the *L*'s and draw lines every three inches along the inside of the short leg. Draw lines every four inches along the inside of the long leg. You don't have to mark the other *L*, because you'll only be using it to cover the marked *L* to create 3:4 rectangles of the size you choose. If you make the adjustable framer from thick cardboard, it will provide years of useful service.

Because of the soft *resolution* of a television screen and the requirements of a good video production, your graphics should feature fairly large bold figures. Pointing a video camera at the small print in this book, for example, wouldn't make a very readable video graphic—the letters are just too small. When you create letters or numbers for a video graphic, the figures should be at least one-quarter inch high for every six inches of TV screen height. For example, if you play your videotape back on a television screen that is twelve inches high, the graphic figures

Figure 6.3 ***An adjustable graphics framer made from thick cardboard.***

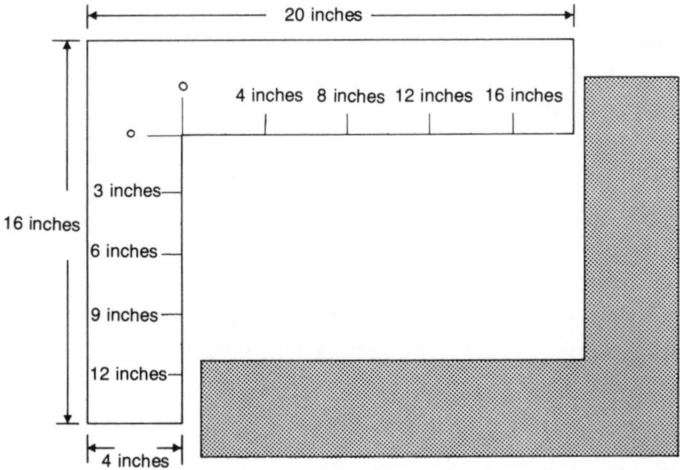

should be no smaller than one-half inch high. You'll have to translate this letter-to-screen size relationship to the dimensions of your graphic frame. If your frame is six inches high, the figures should be no smaller than one-quarter inch; if it is only three inches high, your smallest figure will be one eighth of an inch, and so on.

Graphics in Sequence We mentioned that standard-sized 8½″ × 11″ sheets are convenient for graphic work. Since this is the size used in ordinary three-ring binders, you should have no trouble finding 8½″ × 11″ paper around your home or school. You will also find that ringbinder notebooks are very good for recording a series of separate graphics in sequence.

Let's say that you want to begin your program with four separate graphics. The first one will be the title of the program, the second might indicate a subtitle and the name of the producer or producers, the third will show who directed the program, and the fourth will display the name of one or two major performers.

There is a very simple way to create this graphic sequence. First, you'll need four pieces of firm cardboard and an ordinary three-ring binder. (The cardboard backing from paper pads, poster board, or cardboard cut from manila file folders will do.) Remember, the size of the cardboard sheets should be close to an 8½″ × 11″ standard so that they fit into a ringbinder.

Punch three ringbinder holes along the upper horizontal edge of the cardboard and draw a 6″ × 8″ frame in exactly the same position on each piece of cardboard. You don't want this frame to show, so use a pencil and press lightly. Then, create your letters or graphic figures with felt-tipped markers, rub-off letters, or by any method you choose. Put your finished graphics in the ringbinder in order. Place the ringbinder on some sort of stand, such as an artist's easel or a music stand. Flip the ringbinder cover back so that when you aim a video camera at the stand, it will show the first graphic.

When you begin recording, open with the first graphic (usually by opening the aperture ring from *C* to the proper setting for the prevailing lighting conditions). Then flip the next graphic in the ringbinder over the previous one at whatever speed you think best. Since you placed the graphics in the ringbinder in the correct order, you will be recording them in the correct order. The results will look very neat and professional.

As you create your graphics within the 6″ × 8″ safe area, be sure to line up each one precisely with all the others, so you won't have to move the camera as you flip from one graphic to the next. This kind of graphic-sequencing requires steady camera work, so use a tripod for shooting.

An Extra Note on Graphics for Black-and-White Production If you are producing videotapes in black and white, remember that colors that contrast nicely in color video may not contrast at all in black and white. For example, since red and blue are both fairly dark colors, they will both show up as middle gray on a black-and-white television screen. We've known newcomers to video production who have made the mistake of cutting out graphics from blue construction paper and placing them on a deep red background. To their horror, the result displayed on a TV screen was a uniform gray image. To avoid such a mistake, experiment with some color combinations on your black-and-white camera before you create elaborate graphic materials.

Even if you are producing in color, you should always contrast dark colors with light colors in the design of your graphics. You never know when one of your color programs might have to be played back on a black-and-white TV set.

Before You Try the Following Graphics Techniques

On the following pages, we'll suggest many different ideas for creating video graphics. All these ideas will work, no matter what make or model of portable video system you are using. But the best way to set up your system for any of these techniques will depend on:

- whether your system is color or black and white
- the make and model of your video recorder and camera
- the extra accessories you have (such as special lights)
- available lighting conditions where you do your video work
- your particular tastes and skills

For best results, try out new graphic techniques before using them in an important production. By experimenting with different types of graphics, you can practice and develop some variety in your graphic work. Experimenting will also help you find the graphics methods that best suit your personal style and your particular equipment. Remember, videotape can be used over and over again, so video production lends itself to lots of trial and error.

Simple Do-It-Yourself Graphics

The variety of graphic materials you can prepare for your video productions is limited only by your imagination. Remember, graphics don't have to cost a lot. You can create many interesting and effective graphics from inexpensive materials found in many schools and homes: cardboard, felt-tipped markers, crayons, construction paper, and so on. You can create slightly more advanced graphics from materials available at moderate cost from a good stationery or art supply store: poster board, graphic artists' pens, press-type letters, and so on. If you don't have much experience in creating video graphics, keep two basic points in mind. First, start with the simplest materials; then when you feel comfortable with those, gradually try out more sophisticated techniques. Second, never use expensive materials when cheaper materials would do just as well.

Cardboard and Crayons or Felt-Tipped Pens Cardboard is cheap and easy to use. So are crayons and felt-tipped pens (Magic Markers). These pens are available in a broad variety of colors and with different tip thicknesses (fine point, medium point, wide point).

The cardboard used for video graphics should be firm enough to flip or stand alone without bending. It should be a neutral, fairly light color (light gray and beige are good), so that the cardboard sheets show the true colors of the pens or crayons used to mark them and the markings will show up on a TV screen. You can also use pure white cardboard, but it tends to reflect light and to produce too much contrast with the darker colors of the markers.

Many portable video producers use the rough gray cardboard backing from paper pads as the background for graphics. This type of cardboard creates a rough textured background for graphic work. At the same time, it provides a neutral, nonreflective surface. If this kind of cardboard looks too rough, you can try more expensive poster boards or textured matting boards. Usually, you can buy large sheets of poster or matting board and cut the sheets into pieces that are the right size for video graphics.

Remember that when you work with cardboard and markers, it is a good idea to mark a "safe area" frame (say, 6″ × 8″) lightly in pencil on the cardboard before you draw the graphic. If your graphic is a title or a credit containing several words, measure out your work in advance and draw straight horizontal guidelines in very light pencil. To be even more sure that your graphic will look the way you want, you can pencil a rough sketch of the graphic on a scratch pad before creating your final work. A little extra time in preparation often saves a lot of extra time in correcting mistakes.

The Chalkboard Almost all schools have chalkboards and chalk. Since chalkboards are erasable, chalk graphics allow for much trial and error. Chalk is inexpensive. With so many colors available, you can create chalk graphics in your own special styles and designs. You can also make them at the same time you are videotaping a live performer. For example, you may be in a classroom recording a friend explaining the number of sports injuries in playing soccer as compared with football. While your friend talks, he may want to illustrate his point by writing some figures and statistics on the chalkboard. As your friend writes the statistics on the chalkboard, you could pan and zoom in on what he is writing.

Chalk works just as well for letters, numbers, pictures, charts, and all sorts of graphs. Although chalkboard graphics will never look fully professional, they can be made to look attractive and amusing.

A school chalkboard offers plenty of working space for your artwork. Within this space, however, be sure to outline 3:4 rectangles to define the borders of each new graphic. Draw at camera level (say, four to five feet from the floor). The only real disadvantages of chalkboard graphics are that the chalk is very dusty and it's very hard to record them in sequence.

The Dry-Wipe Board Dry-wipe boards are a new and improved version of the standard chalkboard. A dry-wipe board features a specially treated white surface and special pens that look like felt-tipped markers. Like most markers, dry-wipe markers are available in broad-tip or narrow-tip styles. Figures drawn on a dry-wipe board with these pens can be easily erased, but they do not produce nearly as much dust as chalk. Also, the dry-wipe figures tend to be more crisp and clearly defined than chalk figures, and the white background of these boards usually works better for video graphics than the green or black surface of a chalkboard (although the white surface of the dry-wipe board can sometimes cause the same "glare" problem you run into when you use pure white cardboard as a graphics background).

Unfortunately, dry-wipe boards may not be available in schools or homes. However, they are used in many business meeting rooms and college classrooms, and you, a friend, or your parents might know where you could borrow one. In fact, you may even be able to buy a small one of your own. An 18″ × 24″ dry-wipe board can cost as little as $15.00, and markers are priced in the $1.50–$2.00 range.

You design, draw, and shoot dry-wipe board graphics in much the same manner as chalkboard graphics. And like chalkboard graphics, they are hard to shoot in sequence.

Felt Boards Although felt boards are not standard household items, they are available in many schools. You can easily build one at reasonable cost for home use, too. Homemade felt boards don't have to be large. For video graphics, an outer felt-board frame of 12″ × 16″ (a 3:4 ratio) is big enough. Since felt is an inexpensive cloth with a rough texture, one piece of felt will stick to another very easily.

To make an inexpensive felt board, try stretching a piece of felt tightly over a 12″ × 16″ board or thick piece of cardboard. Then tack down the edges on the other side with thumbtacks or staples. The choice of whether to use light- or dark-colored felt is up to you. It's so easy to do, you might want to do both.

You can use your homemade felt board as a backdrop for a variety of figures cut out from other pieces of felt: letters, numbers, and pictures. Just make sure the letters and figures are cut from a color that contrasts with the color you are using as the background.

You can use your felt board to create charts, graphs, and maps. Besides their low cost and ease of use, they offer another big advantage for video graphics. They have rough dull surfaces so they do not reflect unwanted flashes of light into the camera lens. On the other hand, felt-board graphics share a major disadvantage with chalkboard and dry-wipe graphics—they can't easily be set up to record in sequence.

Rub-Off Letters Rub-off letters usually come in sheets with clear plastic backing. You can create video graphics by rubbing the letters onto paper or cardboard. Rub-off letters are available in many different sizes and styles (or fonts), and they produce very neat, professional-looking results. A typical sheet will have letters, numbers, punctuation marks, and a limited number of other simple graphics. The letters and figures come in either black for a light-colored background or white for dark backgrounds. Two popular brands are Letraset and Prestype.

If you use rub-off letters, you should also use a fairly high grade of backing material, such as a textured matting board or good-quality sketch paper. Using rough cardboard with these letters is a little like using expensive hand-carved wood paneling to cover the inside walls of your garage! Rub-off letters are quite expensive. A standard-sized sheet of letters and numbers costs about $4.50.

Because they have the potential for creating very nice-looking graphics, rub-off letters should be used with care. Working within a 3:4 rectangle, first draw lightly penciled horizontal lines to guide the accurate placement of each letter or number. Be sure to plan for the correct number and size of figures in advance, because once you start rubbing off letters, you can't lift them up and redo them if they're wrong. To correct mistakes, you'll have to start again with new figures on a fresh sheet of backing.

Apply rub-off letters by lining up the letters on the sheet; use a ballpoint pen or dull pencil to transfer the letters with a firm rubbing motion. Be sure to place the sticky side of the sheet down over your background material and try not to press too hard. If you do, you'll crack the letters. On the other hand, if you press too lightly, the letters will tear as you lift the vinyl backing from your graphic. To get a feel for how hard you should rub, it's a good idea to try a few letters on a piece of scrap paper before starting on graphics that you'll actually use in a production.

Since you can use rub-off letters on cardboard or paper backing that you can flip, they lend themselves very nicely to recording a series of graphics in sequence. In fact, when you use rub-off letters, you may want to follow the directions for "Graphics in Sequence" described earlier in this chapter.

Paste-On Vinyl Letters Compared to rub-off letters, paste-on letters are more expensive and available in fewer sizes and fonts, but they are a lot easier to use. Like rub-off letters, paste-on letters are sold in stationery and art supply stores. They too are available in both black and white and are sold in vinyl sheets. With paste-on letters, though, you actually peel the letters from the backing and place them on your paper or cardboard background. This means that you can correct mistakes by lifting off the paste-on letter or figure and reapplying a new one correctly.

Unlike rub-off letters, you don't have to use paste-on letters only on cardboard or paper. Instead, you can stick them to more interesting surfaces, such as clear sheets of plastic (plexiglass) or richly textured pieces of wood or marble. When you use the paste-on letters with cardboard or paper backing, you can easily flip and sequence the graphics. But when you use the letters on fancier surfaces, sequencing is usually a little more difficult.

Slides Ordinary slides can be used to create excellent graphics for video. There are several ways of doing this. One way is to record the slides as you project them onto a hanging screen in a darkened room. When you do this, you may have to adjust your video camera's color balance for the light projected onto the screen. For best results, use a moderately powerful slide projector (at least 500 watts).

Instead of a hanging screen, you might use a rear-screen projector, which is nothing more than a piece of special translucent plastic set at a 45-degree angle from a mirror. The mirror reflects the light coming from the projector onto the back of the screen. Because the screen surface of a rear-screen projector is often fairly small, the light displayed on it is fairly bright, and it doesn't require a darkened room for viewing or for video recording. Since the light qualities and intensities of the projected image will change, you should do some experimenting before seriously attempting to create and record slide graphics for video.

Video graphics made from slides need not come from conventional photographs. You can create interesting effects by writing or drawing on clear slides with narrow-tip markers designed for use on overhead transparencies. For this reason, if you are a photographer you should hang on to your overexposed blank slides. In fact, you can make your own blank slides by cutting rectangles of clear plastic acetate to 1½″ × 1⅜″ and then placing the rectangles in the make-your-own cardboard or plastic slide frames available in most photo supply stores (Fig. 6.4). Acetate is a flexible transparent plastic often used for overhead projector

Figure 6.4
Video graphics made from photographic slides.

transparencies, and you can buy it in sheets or rolls in art supply and school supply stores.

You can also use completely unexposed opaque slides to create video graphics. These are the slides you get back for frames you mistakenly shot before removing your camera's lens cap. When you look at both slides of such a slide, one will be shinier than the other. The dull side contains a black chemical emulsion that can be scraped off its clear plastic base with a sharp blade or a compass point. You can scratch pictures, letters, or numbers on this black emulsion, allowing light to pass through only where you scratched. In some production situations, this kind of "scratched slide" graphic can be useful and appealing.

You should remember two details before trying to create video graphics from slides. First, you will be working on a rather small surface (Fig. 6.5). So if you don't care for painstaking, detailed work, you'd better use some other graphic technique. Second, the aspect ratio for a standard 35 mm slide frame is not 3:4, it is more like 2:3. To allow for

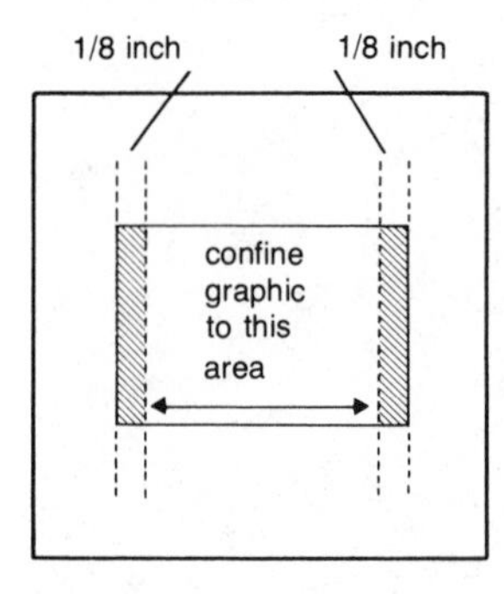

Figure 6.5
Using photo slides to create video graphics.

this when you're creating drawn or scratched slides, try marking a safe-area line about one-eighth inch inside the vertical borders of the frame as shown in Figure 6.5. This will create a 3:4 aspect ratio inside the slide frame—the aspect ratio used in video. In any case, be sure to keep your artwork well inside all four borders of the frame. This will provide maneuvering room when you set up your video camera to capture all the details you created in your slide graphic.

One of the major advantages of a slide graphic is that it allows you to add materials to your video recording that you could not otherwise use (for example, slides from your early childhood and slides of distant places). Also, with an ordinary slide tray and projector (such as a Kodak Carousel), you can very easily sequence your individual graphics in any order.

Animated Graphics

You can create more advanced moving video graphics with materials and supplies found in many homes and schools. These animated graphics tend to be more elaborate than the simpler techniques we have described so far. However, "more elaborate" does not always mean better, so if you can get the job done by using a simple method, think twice before spending time and money on something fancy. The techniques described on the following pages are not really difficult, and each can add interest and appeal to your video programs.

Plexiglass Plexiglass is a clear stiff transparent plastic, usually one-eighth inch thick. You can buy plexiglass at most hardware or glass stores, and you can have it cut to any size you choose. Plexiglass is fairly expensive, costing roughly \$2.00–\$3.00 for a 12″ × 16″ piece. However, you can re-use plexiglass many times. You can also use many different kinds of markers on plexiglass, including:

- felt-tipped markers (the kind used for overhead transparencies)
- permanent felt-tipped markers (''mark anywhere'' markers)
- grease pencils
- vinyl paste-on letters (white or black)

You can easily remove marks made with any of these materials by rubbing the glass with a cloth lightly soaked in water (for overhead transparency markers) or cleaning solvent (for permanent markers and grease pencils). With vinyl paste-on letters, you simply peel the letters off.

You can use plexiglass graphics as a foreground for background motion. For example, you could add program credits to a program by marking the credits on a piece of clear plexiglass and placing the plexiglass against a television screen displaying a video screen recorded earlier. (This would require two video decks: one to play back the tape providing the graphic background and the other to record the plexiglass foreground graphic against the television screen.) You can also place plexiglass graphics in front of still photos, animated puppets, live action, or a slide projected on a rear screen. Plexiglass graphics used in conjunction with electronic video feedback can create spectacular effects. (We'll describe electronic video feedback in just a few pages.)

When you use plexiglass for your graphics, be sure to use dark figures against light backgrounds and vice versa; otherwise your figures will be hard to see. Since plexiglass is transparent, you can't flip one plexiglass graphic over another in sequence, so they usually have to be used one at a time.

If you use a plexiglass graphic against another background, place the plexiglass as close to the background as possible. This way, you'll be able to focus your lens sharply on both the plexiglass and the background at the same time. If they are too far apart, either the plexiglass or the background may be blurred.

Crawling Titles When you watch television or a movie, you very often see the end-of-program credits roll slowly from the bottom of the screen to the top and then out of sight. This is called crawling, or scrolling. The crawling titles you see on TV are usually created with expensive electronic equipment. However, even though you don't have this professional equipment, you can "crawl" titles, too, with the help of a simple homemade device called a title crawl.

You can make a title crawl from readily available lumber and hardware. A crawl is nothing more than a wooden frame with dowels at the top and bottom. The upper dowel rolls with the turn of a crank, so that long strips of paper (or clear acetate) move upward. By aiming a video camera at the graphic material moving between the upper and lower dowels, you can record credits or artwork so they will appear to crawl from the bottom to the top of the television screen. Once you've built the title crawl and created the graphics, all you have to do is turn the crank!

Here is how you make a title crawl. First, you'll need the following parts:

- Two pine two-by-fours, 18 inches long
- One pine two-by-four, 25 inches long
- One pine two-by-four, 16 inches long
- Two pieces of one-by-three pine strapping, each 14 inches long
- One piece of one-by-three pine strapping, roughly 6 inches long
- Two 1¼-inch dowels, each 18¾ inches long
- Four ¼-inch pegs, each 2½ inches long
- Four two-by-four scraps, roughly square in shape
- One wooden spool (from a spool of thread)
- Fourteen 2-inch flat-head wood screws
- Eight 2¼-inch flat-head wood screws
- One 2¼-inch round-head wood screw with washer
- Twelve 2-inch angle irons

You'll also need a screwdriver and a hand-powered or electric drill with 1⅜-inch and ¼-inch drill bits. If you've never used an electric drill before, be sure to get help and instructions from someone who has. In fact, if you haven't had much experience in woodworking, you'll probably need help.

Once you have all your parts and tools together, here's what you do:

Step 1 Drill two 1⅜-inch holes in each of the two 18-inch two-by-fours, roughly 10 inches apart, as shown. The holes in both two-by-fours should line up exactly, because these

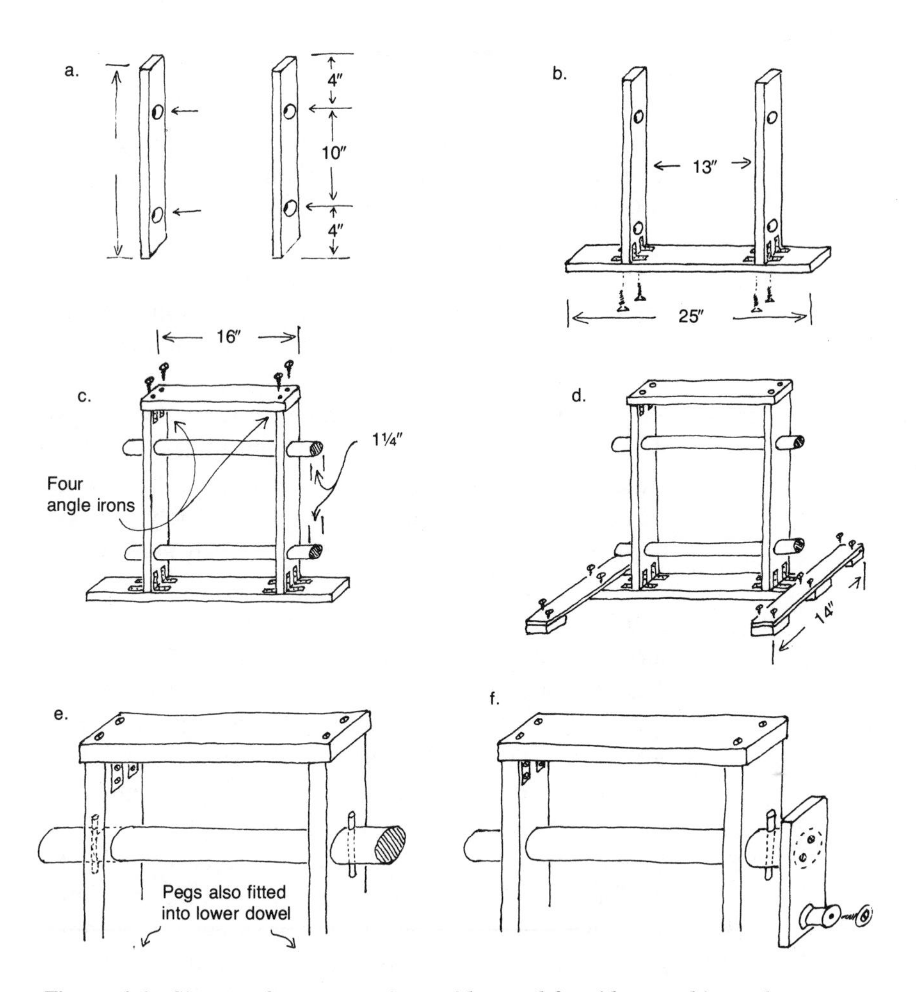

Figure 6.6 *Six steps for constructing a title crawl for video graphic work.*

pieces of wood will become the upright pieces for the crawl frame. The holes will be centered about 4 inches from the ends of these upright pieces.

Step 2 Place the two drilled upright two-by-fours facing each other about 13 inches apart on the 25-inch two-by-four. This 25-inch two-by-four will become the title crawl's base. Secure the base to the upright with eight 2-inch angle irons and four 2¼-inch screws, as shown. You'll have to use the ⅛-inch drill bit to drill starter holes for the screws. You'll also have to "countersink" the holes you've drilled. (To countersink, you drill a very shallow larger hole in the same place that you drilled the starter hole for the screw. This larger hole should be roughly the same size as the screw head. Then when the screw is turned all the way in, its head will not protrude outside the top surface of the wood.)

Step 3 Secure the 16-inch two-by-four to the top of the frame with the remaining four angle irons and the four 2¼-inch flat-head wood screws as shown. This completes the basic frame.

Step 4 Make a solid stand for the frame by screwing the two 14-inch pieces of pine strapping at right angles to the base of the frame, as shown. Use two 2-inch flat-head wood screws at each end. Underneath each end of *both* pieces of strapping, secure a square piece of scrap two-by-four with two 2-inch flat-head wood screws, as shown.

Step 5 Push the 1¼-inch dowels through the upper and lower pairs of holes in the frame's upright pieces, as shown. These become the rollers for the graphic strips. To secure the dowels in place, you will have to drill small quarter-inch holes in them (using the quarter-inch drill bit) about 16½ inches apart. Force the four ¼-inch pegs into these holes, as shown, once the rollers are in place.

Step 6 Make a crank from the 6-inch piece of pine strapping by attaching one end to the upper dowel with the remaining two 2-inch flat-head wood screws as shown. At the other end, attach the thread spool with the round-head 2¼-inch wood screw and washer as shown. Leave the spool loose enough to twirl. It will be the turning handle for the crank.

Use rolls of paper or clear plastic acetate in a title crawl. First, create your graphics on long strips of paper or flexible plastic acetate. Then attach these strips to the crawl. To do this, fasten the bottom edge of your strip to the lower dowel with masking tape. The letters and figures on the graphic strip should be facing right side up and outward. Roll your strip all the way onto the lower dowel. Then pull the top edge of your strip upward to the upper dowel and secure it with masking tape, so that the strip is stretched quite tightly between the two dowels. When you turn the crank clockwise, your figures will "crawl" upward.

With a crawl, there is no need to flip or sequence your graphics, because the crawling action does the sequencing for you. Generally, the title crawl lends itself best to the display of production credits, but you can put the crawl to other uses, too.

One of these uses has nothing to do with graphics. A title crawl can be used instead of cue cards as a *teleprompter* for on-camera performers. In an interview, for instance, you might want to list in bold letters on the crawl all the questions you want the interviewer to ask. With the crawl facing the performers, a production assistant could slowly crawl the questions upward for the interviewer to read as the program moves along.

Earlier, when we described how you can use plexiglass for graphics, we explained how you can place clear plexiglass in front of a TV screen to create animated graphics. You can do the same thing with the crawl, as long as you use clear plastic (rather than paper) for your crawl graphics. Just make sure that the letters, figures, and drawings you place on the plastic crawl strip are in clear contrast to the moving pictures you will be using for background material.

Because plastic acetate is a shiny, flexible material, it tends to reflect annoying flares of light into the camera lens. These flares are generally too short-lived to damage a camera's pickup tube, but they can ruin an otherwise good graphic display. To avoid this, you'll have to experiment with some different placements for the camera, lights, and title crawl. Try to arrange things so the light doesn't reflect off the surface of the acetate.

Electronic Video Feedback When used correctly, *electronic video feedback* can produce spectacular visual effects. Here's how to do it. First, connect your video camera and a television set to a VTR as described in Chapter 2. Switch on the camera, the VTR, and the TV set. Be sure that the TV set's channel selector is set for your VTR (usually Channel 3 or 4). Point the camera at the TV screen. What you'll see is an unusual mix of light and color. When you gradually move and twist the camera, the TV screen will explode into a kaleidoscope of swirling light, patterns, and color. The reason for the spectacular images produced by electronic feedback is the gap between the time the camera picks up an image and the time the TV displays the image. It is impossible to describe feedback adequately with words or still photographs. You'll just have to experience it for yourself.

If you don't use it too often, electronic feedback can add interest and beauty to your video program. Sometimes, you can combine electronic feedback with other visual materials and sound effects to create

Figure 6.7 *Video feedback with dried flowers placed between the video camera and the TV screen.*

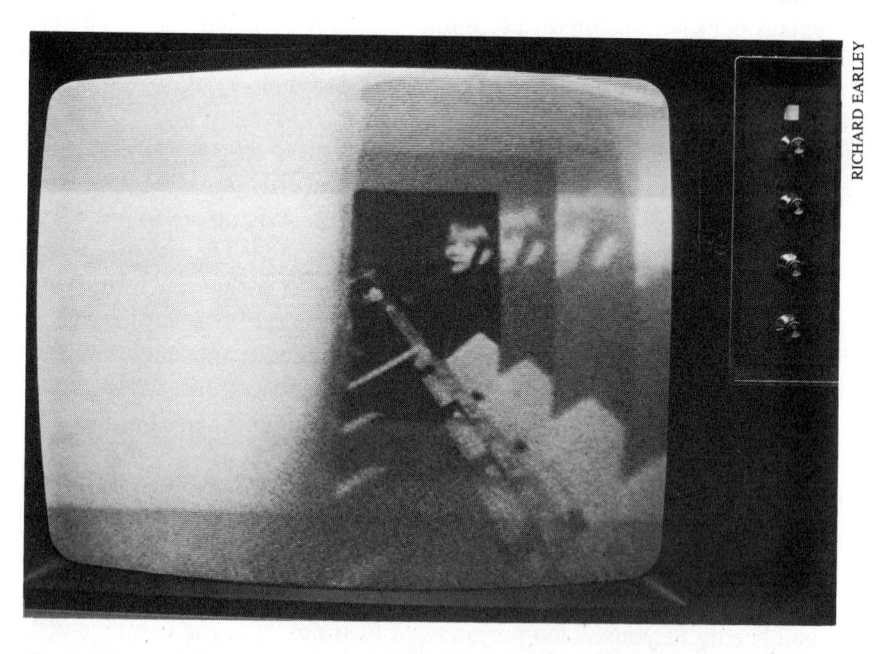

Figure 6.8 *Video feedback with photograph taped to the surface of the TV screen.*

even more interesting effects. For example, if you place a bunch of dried flowers between a television screen and its video camera while you are creating electronic video feedback, the flowers will appear to swirl in and out of the kaleidoscopic feedback images (Fig. 6.7).

In the same way, a still photograph taped to the front of a television screen that is displaying electronic feedback can also produce interesting results (Fig. 6.8).

Since no two feedback images are ever precisely the same, you can never predict exactly what you'll see on the screen. Limit your use of electronic feedback to those few situations where it can make a real contribution to the quality of your program. Otherwise, it can look silly and out of place.

Future Graphics for Homes and Schools

We mentioned earlier that many of the graphics you see in professional television programs are produced electronically. To produce these electronic graphics, professional TV and video studios use a device called a *character generator*. The character generator is a computer with a typewriterlike keyboard that generates letters and numbers (alphanumeric figures) for display on a television screen. These figures are usually mixed with (or placed over) live program material. In other words, the TV screen doesn't go dark when the titles come on—there is still a picture behind the titles. Character generators come in a variety of models, but most models are too expensive for home use. As the market for do-it-yourself video production equipment grows, there is a good possibility that inexpensive character generators will be developed for home use.

Even as we write, Panasonic has announced a new home-type portable color video camera with a built-in character generator. This lightweight consumer model is said to produce good color video in very low light and has remote controls that allow the operator to control the major functions of the VTR deck from the camera itself. If this camera produces a good color picture in addition to the fancy features, it's well worth a close look (Fig. 6.9).

You may have noticed that *microcomputers* (such as those made by Atari, Apple, Radio Shack, Texas Instruments, and other manufacturers) can do many of the things on a television screen that character

PANASONIC

Figure 6.9 ***A new model of video camera that can produce titles electronically (Panasonic PK-802).***

generators do. Can you use it as a character generator? The answer is that as we write (early 1982), the electronic signals produced by most microcomputers are not stable enough to be recorded onto videotape. They *can* feed a viewable signal directly to a TV screen, but the signal *can't* be recorded on tape. However, even now, special *circuit boards* are available for microcomputers that will upgrade the computers' video output signals to the point where they can be recorded and mixed with other video signals. These circuit boards cost more than $200, so for the moment they are out of the price range for most portable video users.

For the future, however, it may be different. Like many other electronic gadgets, the mass-marketing of this kind of circuit board may lower the price drastically. Even more likely, microcomputer manufacturers themselves will probably soon build this kind of circuit board right into the computers they sell. By the time you read this page, this might have happened. In fact, even now the low-priced Sinclair microcomputers put out a signal stable enough to produce characters for video recording.

There is a way to record computer graphics in your video productions right now without any extra circuit boards or upgraded equipment. All you have to do is point your color video camera at a television screen displaying computer graphics. The results will depend on the quality of your video camera and a few other factors, but it may be worth a try.

The day will surely come when you can feed a good computer-generated video signal directly into your portable video recorder. When that happens, the choices you'll have in designing video graphics will expand many times. For example, many of today's microcomputers are able to create many types of graphics. Tomorrow's models will feature even more styles and types of graphics. These new and improved microcomputers will be able to serve as character generators with titles and credits in many different fonts. They also will be able to serve as *word processors* for scripting and other production-related writing tasks. By the mid-1980s, homes and schools with access to microcomputers and portable VTRs will be able to combine these two machines for production activities far more advanced than anything available today.

chapter 7

TAKING IT FROM HERE

Production Possibilities

Now that you know how to use a portable video system, what are you going to do with it? That depends on what you're interested in doing. For example, you may be interested in having a friend or parent videotape you at a basketball or tennis practice-session, so you can analyze your strengths and weaknesses as an athlete. If you're acting in a school play or running for school office, you may also be interested in setting up a camera on a tripod so you can record yourself rehearsing your lines or practicing a campaign speech.

Or you may have more ambitious production plans. For example, you may want to prepare and produce a video documentary (a film or TV program that explores issues in a factual way) on a topic that's important to you. Once you've finished the documentary, you may want to use it as part of a school project, or you may want to have it transmitted over an *access channel* (a cable channel set aside for showing programs produced by the people in a community) on your local cable system.

Here are some different productions you might try. Look them over and see what interests you.

Documentaries on Important Topics In every neighborhood or community, there are issues that affect people of all ages. Often these issues make good topics for video documentaries. Choose a topic that's important to you and plan a documentary that explores all sides of the issue. For example, if you're concerned about the closing of a school in your neighborhood, you might want to produce a documentary that shows how different people and groups feel about the school closing. In your documentary, you might interview parents and students who will be affected by the closing. To be fair, you should also try to interview teachers, school officials, and the school-board or city-council members who voted to close the school. This way, you'll cover all sides of the issue. You might also want to add a video tour of the school. As you show different parts of the school, you could have the school principal explain the history of the building.

If there is a cable TV system in your town, the cable operator might be willing to show your tape on a local channel. Many cable operators welcome productions that explore important community issues.

Family Documentaries Family documentaries can be simple "video home movies," or they can be more complicated productions that require much more planning and preparation. Video home movies are video recordings of family events: trips, weddings, birthday parties, family reunions, and so forth. If you're recording one of these events, you don't usually have to do much advance preparation. Usually, you just set up your portable video system and shoot the action as it happens. Of course, this kind of shooting can be improved with some advance thinking about camera angles, lighting, and so on. Some families videotape important family events and store the recordings in their own videotape library.

Some other family documentaries require much more preproduction preparation. For example, you might want to produce a documentary production that traces your family's history. As a first step, you would probably want to arrange video interviews with older members of your family (grandparents, great-grandparents, great-aunts and -uncles, and so on). What can they tell you about where your family came from and how your family members used to live? You might also want to arrange production trips to homes and locations that were important in your family's history. Finally, you might want to include shots of old family photos, articles, and heirlooms that are stored away in attics or cellars.

Community Histories Community histories are similar to the family histories we just described. No matter where you live, there are always people who know about the history of your city, town, or neighborhood. Arrange interviews with these people and ask them to tell and show you how your community has changed and grown. Using a storyboard or shooting scenario (see Chapter 5), plan a video production that traces your community's history. You'll probably want to include interviews, shots of older buildings, shots that show where important buildings used to be, and shots of old photos and maps taken from books in your town library.

Dramatic Productions Some portable video producers become very skilled at videotaping skits, plays, and other dramatic productions. For starters, you might want to try producing parodies of soap operas, westerns, or some other type of TV drama. (A parody is a funny imitation of a familiar writing or television production style.) For example, a parody of a TV western might include many of the characters you find in a real TV western: a hero in a white hat, a villain in a black hat, and a fast, faithful horse for the hero to ride into the sunset. In the parody, though, the hero might not be all that heroic, the villain might not be all that villainous, and the horse might not be all that fast or faithful. (The scrip for *The Amazing Adventure of Dr. Omega* in Chapter 5 is a parody of a second-rate science fiction TV thriller.)

By watching television closely and producing parodies on what you see, you will gain some experience with the styles and techniques of dramatic video production. Once you have this experience, try writing and recording more serious dramatic productions—dramas that deal with issues and ideas which are important to you. For example, you might want to write and videotape a play about common problems teenagers face at home and school. For this type of production, and most other dramatic productions, you would need to do a lot of research and pre-production preparation. In dramatic productions, it's usually better to work from a detailed script instead of a storyboard or shooting scenario.

Video for Video's Sake

Your video productions don't always have to tell a story or send a message. For example, you might want to produce videotapes whose only purpose is to capture images and sound in an artistic or interesting manner. Or use your video system to experiment with unusual camera shots—shots that could add artistic touches to your documentary and dramatic productions. Here are some ideas to get you started:

What Is That Mystery Form? Use the close-up capacity of your video camera to look closely at the different shapes, patterns, forms, and textures that you find in everyday life—indoors and out. The world is full of interesting and beautiful forms which go unnoticed by most of us in our rush through life. With a video camera, you can freeze some of these forms for observation and reflection. With your friends, you can make a game of this kind of observation. Here's how:

With a friend or a member of your family, take your portable video system to a place where you might find unusual shapes or patterns (the bark of different kinds of trees, an odd mechanical shape on your bicycle, coiled rope, lichens on a rock, and so on). With your camera lens zoom control set at its maximum telephoto setting (or **CLOSE-UP** setting if it has one), record several short segments of tape, say five seconds or so, in the closest detail possible. Be sure that your focus is sharp. Try to find images that would not easily be identified by a television viewer in such close-up detail.

After shooting about ten close-up images, shoot them all over again from far enough away so a viewer could identify the image easily. For example, if your first close-up was a shot of the bark on a strange-looking tree, the longer distance shot would show more of the same tree (Fig. 7.1). Do not rewind your cassette for the second round of shots. You will need both rounds of shooting for playback.

Take your tape back to a small group of friends or family members who didn't participate in the shooting. Play back the first round of close-up images and see how good your viewers are at guessing the mystery forms. Perhaps they can jot down their guesses on pieces of paper. Then play back the second round of shots—the shots that show the more distant views of the same images. See how many times you were able to fool your viewers and how many times they were able to outguess you.

This simple game raises some questions for discussion. For example, how often do professional television producers and film-makers start a new scene with a close-up before showing the broad wide-angle sweep of the scene? Why and when would a producer choose to start with a wide-angle shot (sometimes called an *establishing shot*)? Why and when would the producer start with a close-up?

This game also gives you practice in using your camera in unusual ways. It encourages you to try different camera angles and lens perspectives. To get the detail you want in your first round of shots, you may have to get down on your stomach or climb up onto a table. Be careful with your camera, but don't be afraid to go after the shots you want.

Guess the Mystery Object This activity is very much like the last one. However, instead of recording close-up images of different objects

Figure 7.1a *Tree shot with zoom lens at maximum telephoto setting.*

Figure 7.1b *Tree shot with zoom lens at medium telephoto setting.*

in sequence, you start by recording interesting visual close-ups of a single object, such as a lawn mower. Here you might record three extreme close-ups of different parts of the lawn mower. Then you record three more shots from a *slightly* greater distance. Then, after pulling away a little more, record three more shots. You keep doing this, pulling farther and farther away for each sequence of three shots, until you finally have a shot of the whole lawn mower.

When you play back your tape to a group of friends, ask them to guess the mystery object as soon as they think they know what it is. The first one to guess it correctly is the winner.

A Young Child's View We often think that everybody sees the world exactly as we do. For example, if you are a sixth grader or older, your eyes probably view things from a height of between 4½ and 6 feet. When you're in a crowd of people, if you look straight ahead, you see a lot of other people from the shoulders up. But do you remember what crowds looked like when you were a small child? If you were a small child in a crowd, you would see a confusing forest of pant legs and skirt hems. And, if you weren't holding on to the hand of a trusted adult, you would probably have been frightened.

Use your video camera to recreate what this kind of experience is like. Go out into your schoolyard with a classmate and a portable video system and try to record what you see from a height of two feet off the ground. Try to get a group of friends to create a crowd for you to move through with your camera. Do the same at a basketball game, viewing the game and the crowd from a small child's viewpoint. Feel free to lurch around with the camera to imitate the uncertain, clumsy way that small children walk.

Record your shots from the child's-eye view in three or four different situations. When you've done this, bring the tape back for viewing. Have your viewers guess the point of view that you had in mind as the camera operator. Ask them to suggest other shots you might have taken from the child's viewpoint and other points of view from which another series of shots might be taken (a dog, a very tall basketball player, a house fly, a snake, and so on).

This activity encourages you to look at things with your video camera from unfamiliar points of view. It also gives you practice in using your camera under difficult conditions. When you are producing more serious dramatic or documentary programs, you will need this skill and awareness to shoot your scenes from as great a variety of viewpoints

Figure 7.2a

Figure 7.2b

as possible. Otherwise, your programs will be less interesting and less effective.

Other Ideas There are many other things you can do with your video camera to practice and refine the artistry of your camera work.

You can go out and try to match shapes and patterns in your environment (for example, objects that are triangular or circular in shape). You can also match objects with rhythmic repetitive patterns, such as the stakes on a picket fence or the tines of a pitchfork. You can create sequences of short shots where the pattern of the first relates in some way to the pattern of the second, the pattern of the second relates in another way to the pattern of the third, and so on (Fig. 7.2a–f).

Figure 7.2c

Figure 7.2d

Figure 7.2e

Figure 7.2f

Another activity is to try to create suspense by breaking down an ordinary event into a series of several shots, mixing close-ups with long shots and medium shots. Take the act of getting onto a bicycle, for instance. One way to record this would be to get the camera up on a tripod to get a long shot of a friend mounting a bicycle and riding away. First, you set up the shot. Then, you start your tape rolling and cue your friend to get on his bike and ride out of your camera viewfinder. As soon as your friend moves out of the scene, turn off the VTR. The whole shot should take roughly five or six seconds. Not very interesting, is it? Another way to shoot this scene would be to break the action down into many different parts, each of which would be its own separate shot, as in Figure 7.3 (Shots 1–9).

To record these shots in sequence, follow the directions for stop-and-start transitions found in Chapter 4. For each new shot, you'll have to move your camera and set it up again. As you record the shots, you'll be creating an artificial sequence of events. But your viewers will see these events as real, not artificial, just as they do when they watch TV or go to a movie.

This kind of artificial shot-sequencing is used all the time by professionals. If the viewer bothers to think about it, he knows that the recorded sequence of events is not real, that getting onto a bicycle would not really happen in the manner shown. But the viewer chooses not to think about it because the event is shown in an interesting way. This is called the viewer's "willing suspension of disbelief."

When events are recorded in this detailed way, they are also stretched out in time. Each of the nine shots suggested above would take about three or four seconds. So an event that would take only five or six seconds in real life now takes about thirty seconds when shown in

Figure 7.3a

SHOT 1 (LS) Friend approaches bicycle, reaches for handlebar.

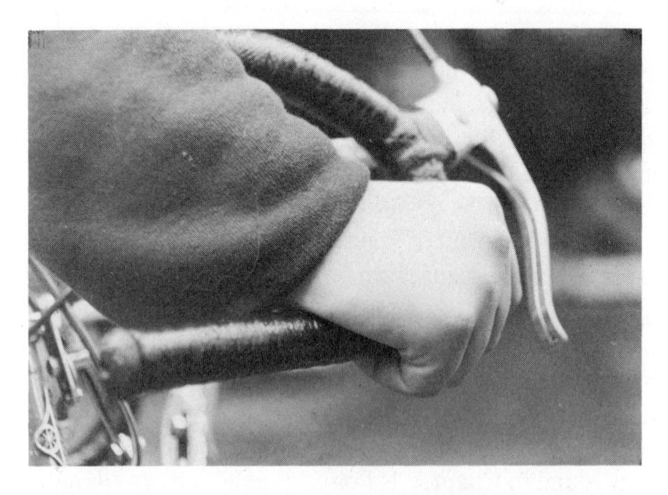

Figure 7.3b

SHOT 2 (CU) Friend's hand grips handlebar.

Figure 7.3c

SHOT 3 (CU) Friend's right foot steps on pedal.

Figure 7.3d

SHOT 4 (MS) Friend's left foot swings over the bicycle's seat.

Figure 7.3e

SHOT 5 (LS) Friend settles on bicycle, adjusts the gears.

Figure 7.3f

SHOT 6 (CU) Friend tests one of the hand brakes.

Figure 7.3g

SHOT 7 (CU) Friend pushes off with left foot.

Figure 7.3h

SHOT 8 (MS—pan right to follow action) Friend's feet are shown starting to pedal, picking up speed.

Figure 7.3i

SHOT 9 (LS) Friend speeds away out of the camera's viewfinder.

shot-by-shot detail—about five or six times as long. Film-makers often call this "cinematic time." The use of cinematic time is intentional, it improves a production, and it is very different from real time. There are many ways for you to use your video camera to explore cinematic time. Think of some others and try them out.

For more activities that can help you develop your skills as a video artist, look over Chapter 3 of John LeBaron's book, *Making Television.** The book *Doing the Media*** by Kit Laybourne and Pauline Cianciolo also has many good ideas.

For another interesting video art effect, try experimenting with video feedback. Chapter 6 includes directions for creating video feedback and suggestions for ways you might use video feedback in your productions.

Although we've listed some suggestions here, the most important source of video production ideas is always your own mind. Always begin by deciding what's important to you and how you might express your ideas and concerns in a video production.

Some Important Questions for Video Production

In choosing one type of production over another, you should ask yourself a few key questions. We've listed four of those questions here:

1. *What is my purpose in shooting?* Because any kind of video production takes a fair amount of time and effort, you should have some idea of why you're shooting before taking the trouble to set up your equipment. Your reason can be as simple as getting more practice with your camera work or as complex as doing scripted video drama. If you think through an idea beforehand, you'll have something to guide and organize your shooting activity.
2. *How much time do I have?* Documentary and dramatic productions usually require much more time and preparation than video home movie productions. If you are thinking of produc-

*New York: Columbia Teacher's College Press, 1981.
**New York: McGraw-Hill, 1978.

ing a documentary or drama, set aside enough time to do the planning and the production.

3. *What kind of pre-production activities will work best for the type of production I've chosen?* We described three different pre-production activities in Chapter 5: shooting scenarios, scripts, and storyboards. Scenarios usually take less time, while scripts and storyboards take more time. If you've decided to produce a drama, you may need a detailed script. A documentary might require a shooting scenario. For video art productions or video home movies, you may not need either a scenario or storyboard.
4. *How much help will I need?* Dramatic and documentary productions usually require crews of at least three people: a director, a camera/VTR operator, and a sound assistant, not to mention the people you'll be taping. Dramatic productions also require actors and actresses. If you are planning a documentary or dramatic production, make sure you have enough people on hand to do the job. The producer's checklist described in Chapter 5 will help you do this. The checklist will also help you make sure that you have all the equipment and materials you'll need to complete your work.

Showing and Sharing Your Tapes

What should you do with a completed video production? That depends on the type of production and the audience for your program. For example, if the production is a video home movie of a family reunion, you may not be interested in showing it to anyone but family members. Besides, no one else may be interested in seeing it!

But there will be some tapes that you would like to show to other people and that others would like to see. You may also be interested in seeing the videotapes produced by other young people, so you can compare and exchange ideas and production techniques. Here are some suggestions for showing and sharing your video productions.

Local Tape Libraries Most schools have a media center or a library, and most towns and cities have at least one public library. In recent years, some school and public libraries have added a collection of videocassettes to their collections of books, magazines, and records. If your school or public library has a videotape collection, the librarian might be willing to add your videocassettes and those of other young producers to the collection.

If your school or public library doesn't have a videocassette collection, you may be able to convince the librarian to start one. Or you and your friends may be able to start your own videocassette library, using a separate room or a small corner of your school or public library. The librarian may be able to help you organize your collection. You'll need to set up a way to keep track of the video cassettes, and rules for borrowing and returning them. You'll also need to come up with ways of advertising your new service and methods for encouraging other video producers to add their cassettes to the collection. Local school newsletters and community newspapers may help you do this—and give your cassette library some free publicity too!

Starting Your Own Tape-Exchange Group To start a tape-exchange group, all you need is a group of people interested in producing and sharing videotapes on a regular schedule. The members of the group don't even have to be from the same town or city. Using the mail, you can set up a videocassette exchange with people from other places. Sharing tapes with people in other locations is a good way to learn about where they live and what they do. This type of video pen-pal arrangement can even start some long-lasting friendships.

How do you find people who are interested in starting a tape-exchange group? First, you might try posting a notice on bulletin boards at your school, or on the "electronic bulletin boards" offered over some cable systems. You might also visit video equipment stores in your community. Ask the store owners if they know of other young people who are involved in video production and if you could post notices in their stores. If you are forming the tape exchange as a school club, a teacher or school librarian/media specialist might help you get started.

If the people in your tape-exchange group are from the same town or school, you might want to hold regular meetings. At the meetings, you can swap and watch tapes and help each other out by suggesting production ideas and techniques. If your tape-exchange group has been formed as a school club, you might also want to hold fund-raisers to earn money for supplies, trips, and activities. Be sure to use bulletin boards, community newspapers, and school newsletters to advertise your meetings and activities.

Cable Television If your town has cable television, you may be able to show your tapes over a cable channel that is set aside for public or educational use. In towns that have these public-access channels, anyone who has a tape to show can arrange for time on a channel. The cable system may also have equipment you can use to record and edit your public-access productions. In some towns, young people have organized themselves to "cablecast" their own regular news and entertainment programs on local cable channels. Some of these groups have been organized through the public school systems. Other groups of young people have organized themselves.

Check to see what sorts of public-access channels and equipment are available in your community and what rules and procedures you should follow to use the channels and equipment. Because they reach anyone in the town who subscribes to cable, access channels are an especially good place to show community histories and documentaries you've produced. You should also check to see if the schools in your town are equipped to send programs out over the cable system. If they are, a teacher or school media specialist may let you use your school's cable connection to send programs out on a cable channel.

Once you've arranged to show a production on your local cable system, be sure to arrange some publicity for the cablecast. Try placing ads or announcements in local newspapers and school newspapers. The announcements should explain what your production is about, when it will be cablecast, and what cable channel it will be on. You could also try talking to people who work for the cable system. See if they could include announcements about your production in the program schedules or with the monthly bills mailed to cable subscribers.

Some Final Thoughts

Like most video producers, you'll probably find that each program you produce raises as many questions as it answers. After each production, you may ask yourself why certain shots turned out the way they did and why some production techniques seemed to work while others didn't. You may also ask yourself how you can adapt different production styles and forms to fit your own developing needs and interests as a video producer. Asking questions of this kind is a good sign, because it shows that you are exploring the capabilities of your video equipment and experimenting with different production techniques to find out which work best for you.

How do you get answers to your questions? First, you might try reading some of the books or writing to some of the organizations we've described in this section. You should also call or visit video equipment stores in your community—especially the store that originally sold the portable video system you are using. Ask your school media specialist or AV technician for help if you need it. The staff at local TV stations and your local cable system may also be able to answer questions you have. In fact, a tour of a local TV station or cable studio is a good way to learn about the equipment and production techniques professional TV producers use. Of course, your friends and family may be good sources of help and information, too. Finally, we hope that you will think of this book as an important accessory to your portable video system, and that you will come back to these pages for information and ideas that can help you grow and develop as a video producer.

Getting More Information

In writing this book, we've tried to give you all the information you need to plan and produce basic video productions. On the following pages, we describe books and organizations that can supply you with extra information.

bibliography

Barnouw, Erik. *Tube of Plenty*. New York: Oxford University Press, 1975. Information on the history of TV and radio in the United States.

Brown, Les. *Les Brown's Encyclopedia of Television*. New York: Zoetrope Publishers, 1982. Information on the history of television, plus facts and figures on TV programming and the TV industry.

Kane, Heidi, and Debi Bilowit, eds. *Critical Television Viewing*. New York: Cambridge/The New York Times, 1980. Suggestions on how to view TV more carefully and critically. Written for elementary school students.

Laybourne, Kit, and Pauline Cianciolo, eds. *Doing the Media*. New York: McGraw-Hill, 1978. Information and project ideas to help you understand how the mass media work (and how they work on you). Good ideas on how to create your own film, slide, TV, and video productions.

LeBaron, John. *Making Television: A Video Production Guide for Teachers*. New York: Columbia Teachers College Press, 1981. Information on more advanced production techniques and projects.

Polk, Lee, and Eda LeShan. *The Incredible Television Machine*. New York: Macmillan, 1977. Explains the way television and the TV industry work.

Schrank, Jeffrey, ed. *The TV Action Book*. Evanston, Illinois: McDougal Littell, 1974. Activities to help you keep track of your TV viewing habits and to collect information about TV programming and advertising.

White, Ned. *Inside Television: A Guide to Critical Viewing*. Palo Alto, California: Science and Behavior Books, 1980. Information and suggestions to help you watch TV more carefully and critically. Written for high school students.

Wrighter, Carl P. *I Can Sell You Anything*. New York: Random House, 1972. Explains the tricks TV advertisers use to sell their products.

organizations to write to:

Action for Children's Television, 46 Austin St., Newtonville, MA 02160. To learn how ACT is working for better young people's programming on the broadcast networks and cable TV.

Cable Television Information Center, 1800 N. Kent St., Suite 1007, Arlington, VA 22209. To request information about cable TV and to learn how young people are using cable TV in their communities.

Federal Communications Commission, Washington, DC 20580. To request information about television rules and regulations and to request FCC action when broadcasters and cable operators have broken the rules.

Federal Trade Commission, 26 Federal Plaza, Washington, DC 20580. To request information about the FTC's role in regulating TV advertising and to complain about unfair or deceptive TV advertising.

National Federation of Local Cable Programmers, 906 Pennsylvania Ave., S.E., Washington, DC 20002. To request information about cable TV and to learn how you can use local cable channels.

Public Broadcasting Service, 485 L'Enfant Plaza, Washington, DC 20024. To request information about and to suggest ideas for public TV programming.

Television Information Office, 745 Fifth Ave., New York, NY 10019. To request information about programming on the commercial TV networks and materials describing how the networks produce and distribute programming.

WNET/Thirteen, Education Division, 356 W. 58th St., New York, NY 10019. To request information about WNET's "critical viewing skills" project designed to teach young people to watch TV more carefully and critically.

To request information about network programming and to express your opinions about programming or advertising practices write to the TV networks:

American Broadcasting Corporation, 1330 Avenue of the Americas, New York, NY 10019

Columbia Broadcasting System, 51 W. 52nd St., New York, NY 10019

National Broadcasting Company, 30 Rockefeller Plaza, New York, NY 10020

glossary

(Italicized words are defined elsewhere in the glossary.)

AC Adapter The *component* in a portable video system that changes regular house current (*AC power*) to battery-type current (*DC power*).

AC (alternating current) Power The type of power available from ordinary electrical wall outlets. Compare to *DC power*.

Alphanumeric A video *graphic* that contains letters and numbers. Compare with *pictorial*.

Aperture The size of the lens opening created by the *iris* of a camera lens.

Artificial Light Any light source other than the sun. Fluorescent lights, ordinary bulb-type lights, and special TV lights are all artificial light sources.

Aspect Ratio The standard ratio of height to width in any displayed *image*.

Audio Dub Replacing the *audio track* on a segment of *videotape* without changing the *video track*.

Audio Head A stationary *electromagnet* that records and plays back sounds (audio signals) onto the videotape.

Audio Line Source A record player, an audio tape recorder, a second VTR, or any other component (other than a *microphone*) that sends an audio signal into a VTR deck.

Audio Track The strip on the *videotape* where sounds (audio signals) are recorded.

Automatic Gain Control An electronic circuit inside a video camera or VTR that automatically adjusts the *levels* (or volume) of the audio or video signals during a video recording.

Automatic Iris Control A feature found on many portable video cameras that adjusts the *aperture* opening automatically, so the correct amount of light enters the camera. See *iris*.

Back Light An *artificial light* that is placed above and behind one *subject*.

Beta Format System The type of home videocassette recording system invented by Sony and manufactured by Sony, Sanyo, Toshiba, and Zenith.

Built-in Microphone A microphone that is attached or ''built into'' a portable video *camera*.

Burn A scar on the camera's *pickup tube* caused by pointing the camera at a bright light.

Cable Two or more electric wires grouped together and covered with a protective plastic coating. See *coaxial cable*.

Cable Television A service that receives TV signals from various local, regional, and national sources and relays those signals for a fee to subscribers through special cables.

Camera Body The main part of the camera where the *pickup tube* and most of the camera's electronic circuits are housed.

Camera Operator The *crew* member responsible for working the camera in a video recording session.

Capstan The pinch roller in a VTR that controls the speed of the videotape during playback and recording.

Cardioid Microphone A microphone with a heart-shaped *pickup pattern*.

Cassette A plastic casing that holds *magnetic tape* and two plastic reels.

Cathode Ray Tube (CRT) The type of *picture tube* used in most TV sets.

Character Generator A typewriterlike machine used to display numbers, letters, and other *graphic* designs on a television screen.

Charge Couple Device (CCD) Camera A new type of video camera that uses a light-sensitive CCD chip instead of a *pickup tube*.

Chrominance The color information in a *composite color video signal*. Compare with *luminance*.

Close-up A camera shot that shows a narrow area of a scene.

Coaxial Cable A type of *cable* that is especially well suited for carrying TV signals.

Color Temperature The measurement of the exact color quality of a light source.

Component Any complete piece of audio or video equipment.

Composite Color Video Signal An electronic signal that contains *luminance* (brightness), *chrominance* (color), and *synchronization* (sync) information.

Condenser Microphone A high quality microphone that requires its own battery or AC power supply.

Connector A *plug* or *jack* found in a video component or at the end of a connecting cable. See also *input* and *output*.

Control Rings Rings on the camera *lens barrel* used to adjust the *aperture*, focus, and *zoom* settings.

Crawl Words, numbers, or *titles* that move (crawl) across a TV screen.

Crew A video production team, headed by the *director*.

Crop To cut out parts of a scene by moving the camera or adjusting the camera's zoom ring.

Cue 1. To pause a record or tape at the beginning of a segment that will be added to a video production. 2. A *director*'s command that signals the performers to begin performing.

Cut **1.** Switching directly from one scene or shot to the next. 2. A *director*'s command to stop recording and end the scene abruptly.

DC Power Direct current electric power—the type of power produced by batteries and *AC adapters*.

Deck The *component* in a video system that holds the videocassette during recording or playback. Also houses the various electromagnetic heads, tape *transport mechanism*, and the video recording and playback circuits.

Director The person in charge of a video recording session. Compare to *producer*.

Dolly 1. A set of wheels that can be attached to the legs of a tripod. 2. To move a video camera toward the subject (dolly in) or away from the subject (dolly out).

Drop Outs Streaks or spots in a displayed video picture caused by a worn videotape.

Dynamic Microphone A type of *microphone* that does not require its own battery or power supply—generally more rugged than *condenser mikes*.

Editing Rearranging and piecing together shots and scenes to produce a finished video program.

Electromagnet A strip of metal wrapped with a coil of wire that becomes magnetized when electricity is passed through the wire. *Audio heads, erase heads,* and *video heads* are electromagnets.

Electron Gun A device that releases a stream of electrons known as the *scanning beam*.

Electronic Video Editing A way of creating finished video programs with two specially designed VTRs: a playback deck and a record deck. The playback deck plays back shots and scenes recorded earlier and the record deck records the shots and scenes in the order you want for the finished (or edited) video program. See *master tape*.

Electronic Viewfinder A video camera *viewfinder* that is actually a small television screen that displays the *image* being picked up by the camera lens and can also play back images sent from the tape in a VTR. Compare with *optical viewfinder* and *through-the-lens viewfinder*.

Erase Head An *electromagnet* inside a VTR that removes (erases) all prerecorded signals from a videotape so that new signals can be recorded in their place.

Establishing Shot A *long shot* that shows the location of the opening scene.

External Microphone Any *microphone* that is not attached to or built into the *camera body*.

Extreme Close-Up A camera shot that fills the screen with a small section of the subject in very close detail.

Fade To gradually increase (fade up) or decrease (fade down) the strength of an audio or video signal. When a TV picture fades up, the picture gradually appears on the screen. When a TV picture fades down, the picture gradually disappears from the screen.

Ferrous Oxide Tiny metallic particles that coat one side of videotapes and audiotapes. During video recording, the ferrous oxide particles are magnetized into patterns that correspond to the electronic *audio, video,* and *sync* signals.

Field Half a video *frame*. To create a field, the electron *scanning beam* traces every other line from the top to the bottom of the *target* or *raster* area.

Fill Light A light that softens or "fills in" shadows created by *key lights* and *back lights*.

Focal Length The distance between the central optical element in a video camera's lens and the *target area* of the pickup tube. The *zoom* control on some lenses varies the focal length to make subjects look closer or farther away from the camera.

Format The word that designates a type or kind of videotape recorder. The format of a VTR is determined by a number of factors, including the width of the tape used and the *tape path* that the tape follows inside the deck. *Beta, VHS, U-type,* and *Quadruplex* are all different format video recorders.

Frame One complete video picture. One frame includes two *fields*, or 525 horizontal *lines* of video information.

Framing The positioning and arrangement of objects within the borders of a video *image*.

Freeze Frame See *still frame*.

F-Stops Numbers that measure the size of the lens opening. Video cameras with an *aperture* control ring have the F-stop numbers marked on the *lens barrel*.

Function Controls The tape *transport* buttons (play, stop, record, and so on) found on all portable video decks.

Gain The strength (or *level*) of an electronic audio or video signal. ''Turning up the gain'' means boosting the level of an audio or video gain control. See *automatic gain control*.

Glitch Any distortion or interruption in the video picture.

Graphics Any words, still photographs, charts, maps or other printed material that appears on the screen during a television program. See *alphanumeric* and *pictorial*.

Head A small, delicate electromagnet located along the *tape path* inside a VTR. Different heads perform different jobs during the video recording and playback process. See *audio head, erase head, sync head*, and *video head*.

Headroom The amount of space between the top of the subject's head and the top of the picture screen.

Helical scan The system for recording video signals used in nearly all small-format VTRs. Two or more video *heads* are mounted opposite each other on a rotating drum. The heads spin against the videotape as the tape moves in the opposite direction, placing slanted strips of video information on the tape.

High-Angle Shot A shot taken from above the subject, looking down. High-angle shots tend to make the subject look smaller and less important.

Horizontal Sync An electronic pulse that controls the rate at which the *scanning beam* travels across each *line*. Compare to *vertical sync*.

Image The picture that appears on a TV screen or video camera *viewfinder*.

Impedance The resistance of an electronic component or cable to the flow of electric current. All *microphones* have a high or low impedance rating.

In-Camera Editing Editing a video production by pausing the tape between shots or scenes.

Inches Per Second (IPS) A measure of videotape speed as it passes over the *heads* of a VTR. For example, when a VHS videotape is played at standard speed, the tape moves past the video heads at 1.32 inches per second. See *tape-to-head speed*.

Input 1. A *jack* on a video *component* that accepts signals from another component. 2. Any *audio* or *video signal* that is sent into a video component. Compare to *output*.

Iris The adjustable metal leaves inside a camera *lens* that open or close to adjust the *aperture*, allowing more or less light into the camera.

Jack Any *input* or *output* on a video component. Jacks are the receptacles for the *plugs* that are found at the end of connecting cables.

Key Light A main light that is focused directly onto a subject during an indoor video recording session.

Kinescope An outdated film process for recording television programs.

Large-Format VTR A videotape recorder that uses tape which is more than one inch wide.

Lens In a video camera, a series of curved glass pieces, called elements, that focus light onto the *target area* of the camera's *pickup tube*. When you *focus* or *zoom* a camera, you change the distance between the lens elements inside the *lens barrel*.

Lens Barrel The metal or plastic tube on the front of the camera that holds the glass lens elements. The lens barrels on most video cameras have *control rings* for adjusting the *focus, zoom,* and *aperture* settings.

Lens Filter A circular piece of glass or plastic that screws or snaps onto the end of a camera lens. In video, lens filters can be used to adjust the camera for unusual lighting conditions or to create simple *special effects*.

Lens Filter Control The control that adjusts a video camera for changing light conditions. Lens filter controls generally have settings for bright sunlight, cloudy sunlight, and indoor lighting.

Lens Opening See *aperture, iris*.

Level The measurement of the *gain*, or strength, of a signal. The higher the level, the stronger the signal.

Line (of video information) The horizontal path followed by the *scanning beam* as it traces a picture onto a TV screen.

Line-In Jacks The *jacks* on a VTR that are designed to accept audio and video signals from another audio or video *component*. See *input* and *output*.

Long Shot A camera shot that shows a wide area of a scene.

Low-Angle Shot A shot taken with the camera located below the subject, looking up. Low-angle shots make the subject appear large and important to the viewer.

Luminance The black-and-white (or brightness) information contained in a color video signal.

Magnetic Tape A long, thin strip of *mylar* with magnetically sensitive *ferrous oxide* particles bonded to one side. Magnetic tapes are used for recording video and audio signals and also for storing computer information. See *videotape*.

Master Tape The original copy of a finished, or edited, video program.

Matching Transformer A device that allows you to connect electronic components having different *impedances*.

Maximum Telephoto Setting The *zoom lens* setting that makes the subject look as close as possible. Compare to *wide-angle setting*.

Medium Shot A shot midway between a *close-up* and a *long shot*.

Microphone The *component* that changes sound waves into electronic signals. See *dynamic microphone, condenser microphone,* and *pickup pattern*.

Miniplug A small *plug* found at the end of many microphone, audio, and remote control cables, also called a mini phone plug. Compare to *phone plug*.

Monitor A special TV that can accept and display audio and video signals sent directly from a VTR, but that cannot receive normal TV broadcast signals.

Monitor/Receiver A TV set that can act as both a *monitor* and a TV receiver.

Mylar A thin strong plastic substance that is used as a base for *videotapes* and other *magnetic tapes*.

Normal-Angle Shot A shot taken with the camera in front of the subjects, positioned at about eye level.

Omnidirectional A microphone *pickup pattern*. Omnidirectional mikes are sensitive to sound approaching from all directions. Compare to *cardioid* and *unidirectional*.

Open-Reel VTR The type of videotape recorder that must be threaded by hand and winds the tape on exposed reels. Compare to *VCR*.

Optical Viewfinder An inexpensive peep-type *viewfinder* used on some portable video cameras. An optical viewfinder is a separate lens, mounted on the top or to the side of the camera. Compare to *electronic viewfinder* and *through-the-lens viewfinder*.

Output A *jack* on a video component through which signals are sent out to another component. Compare to *input*.

Pan To pivot the camera to the left (pan left) or right (pan right) to follow the action in a video recording session.

Pass One complete journey of a videotape through a VTR either while recording, playing back, rewinding, or fast forwarding. "Pass" can also refer to one sweep of a video head across the videotape.

Phone Plug A large metal *plug* sometimes used for connecting headphones or microphones to a VTR. Also called a quarter-inch plug. Compare to *miniplug*.

Phosphor Dots The chemical coating on the *raster area* of a TV picture tube that glows to create a television picture. In a color television set, the phosphor coating is made up of red, green, and blue dots. See *primary colors*.

Photosensitive Surface Any surface that responds to changes in light intensity. In a video camera, the *target area* of the *pickup tube* is the photosensitive surface.

Pickup Pattern The direction(s) from which a *microphone* will pick up sound. See *cardioid, omnidirectional*, and *unidirectional*.

Pickup Tube The tube inside a video camera that changes light entering the camera into an electronic video signal. The main parts of a pickup tube are the *target area* and the *electron gun*.

Pictorial A picture, drawing, chart, or still photograph used in a video *graphic*. Compare to *alphanumeric*.

Picture Tube The device inside a TV set or monitor that creates an image from the electronic video signal. The main parts of a picture tube are the *raster area* (picture screen) and the *electron gun*. Also called *cathode ray tube*.

Plug A metal device at the end of a connecting cable that plugs into a *jack* on a video *component*. See *input, output, miniplug, phone plug*, and *RCA plug*.

Pop Filter A foam, metal, or plastic shield that fits over the top of a microphone. Pop filters help block out the "pops" that occur when a performer speaks too close to the mike and the annoying rumbling sound sometimes caused by wind at an outdoor recording location. Sometimes called a wind filter.

Portapak Originally a portable half-inch *open-reel* video system. Many video producers also use "portapak" to refer to newer *Beta, VHS*, and *U-Type* portable video systems.

Power Zoom A motorized *zoom* control found on some portable video cameras.

Primary Colors The three colors that combine to create all other colors. In color video, the three primary colors are red, green, and blue. A color video system mixes these colors to create all the shades you see on a color TV screen.

Producer The person who organizes and manages all aspects of a TV or video production.

Production Any TV or video program, or the process of putting together a program. "Production" often refers only to the actual video recording session.

Profile Angle Shot A shot that shows subjects from a side view.

Public Access A *cable television* channel reserved for programming produced by people within a community.

Quadruplex (or Quad) VTR An expensive *large-format VTR* that uses four *video heads* and that is used by many TV stations and professional TV studios. Compare to *helical-scan VTR*.

Raster Area The inside surface of a TV *picture tube* screen that displays the television image. The raster area is coated with thousands of *phosphor dots* that glow in response to the strength of a *scanning beam*.

RCA Plug A small *connector* found at the end of some connecting cables. RCA plugs are usually found on connecting cables that carry *audio signals*.

Record/Standby Mode See *standby*.

Resolution The amount of detail that is present in a film or video image. High-resolution images are very sharp and show very fine details. Low-resolution images are fuzzier and show less detail.

RF (Radio Frequency) The range of electronic frequencies used for transmitting radio and television signals. Television sets are designed to receive programming on particular radio frequencies called channels.

RF Cable In video recording, a connecting cable that carries the *RF* signal from the *deck* to the TV set.

RF Converter A small device inside many portable VTRs that combines the separate audio, video, and sync signals into one low-level *RF* (*Radio Frequency*) signal. This RF signal then travels through a connecting cable to a TV set, where it can be received on a regular television channel.

Safe Title Area The central 80–85% of a television *graphic* where words and numbers may be drawn without the danger of their touching or being cut off by the edges of the TV screen.

Scanning Beam The focused stream of electrons released by the *electron gun*. In a TV *picture tube*, the scanning beam traces an image onto the *raster area*. In a video camera's *pickup tube*, the scanning beam ''reads'' the visual image that the camera lens has focused on the *target area*.

Scenario A rough written outline of the actions, sounds, and scenes that will make up a video production. Compare to *script* and *storyboard*.

Scene A section of a TV or video program. In particular, a sequence of actions that takes place through a single shooting session. Also, the location at which the action occurs.

Script A detailed shot-by-shot outline of a video production. Compare to *scenario* and *storyboard*.

Shot A single video image. See *long shot*, *medium shot*, *close-up*, and *extreme close-up*.

Slant Track See *helical scan*.

Small-Format VTR Any VTR that uses tape which is one inch wide or less. *Beta* and *VHS* are two types of small-format videocassette recorders designed for home use. See also *format*.

Special Effects Unusual images or image *transitions* that require special equipment or special camera techniques.

Special Effects Generator (SEG) The *component* that coordinates the cameras (in a multi-camera production) and creates various *special effects*.

Special Effects Lens A *lens* that alters or distorts the video image to create a desired effect.

Standby 1. A *director*'s command indicating that a recording session is about to start. 2. The mode that a VTR is in when you ''pause'' the tape while recording. In the standby mode, there will be an image on the camera's *electronic viewfinder*, and the deck will begin recording again as soon as you disengage the pause control.

Still Frame A playback effect in which one frame of a video image appears frozen on the screen.

Storyboard A series of sketches that outline how the shots in a video production will look.

Subject The main point of interest in a video image.

Sync (synchronization) Head An electromagnet inside a VTR that records and plays back *sync* (synchronization) signals on a videotape. Also called control head.

Sync (synchronization) Signal Electric pulses that time the movements of the *scanning beam*. In portable home video systems, the sync signal is generated by the camera, recorded onto the videotape, and then used by the TV *picture tube* to recreate the image correctly. The sync signal is made up of *horizontal sync* pulses and *vertical sync* pulses.

Sync (synchronization) Track The section of the videotape where *synchronization (sync) signals* are recorded. Sync signals are recorded along the bottom edge of *Beta* and *VHS* format videotapes.

Talent Anyone who appears in front of the camera during a video recording session.

Tape Path The path followed by the videotape during video recording or playback. The *video*, *audio*, *sync*, and *erase heads* are located at different points along the tape path. Tape path is one of the factors that determines a VTR's *format*.

Tape-to-Head Speed The speed at which the tape travels during playback or recording, *plus* the speed at which the video *heads* spin. See also *helical scan* and *quadruplex VTR*.

Target Area The *photosensitive* front service of a camera's *pickup tube*.

Tear A horizontal line of distortion that sometimes appears across a video image during *playback*. See *tracking*.

Telephoto Lens A camera lens that acts as a telescope, making distant objects appear closer. Compare to *wide-angle lens*. See *zoom lens*.

Three-Point Lighting Arrangement A standard method of arranging *artificial lights*, using a *key light*, a *back light*, and a *fill light*.

Through-the-Lens (TTL) Viewfinder A video camera *viewfinder* that allows the camera operator to look through the camera lens to see the exact image that the camera is focusing on the *target area*. Compare with *electronic viewfinder* and *optical viewfinder*.

Tilt Pivoting the camera up (tilt up) or down (tilt down) to follow action or frame a shot. Compare to *pan*.

Titles *Credits*, captions, or other printed information recorded as part of a video production. See also *graphic*.

Tracking The position of the tape as it passes over the *video heads*. If the tape is not tracking correctly, a *tear* may appear in the TV screen during playback.

Transition In video, a way of moving from one *shot* or *scene* to the next.

Transport (mechanism) The wheels, motors, and other mechanical parts that thread and move the videotape through a VTR. See *capstan* and *function controls*.

Truck To move the camera to the left (truck left) or right (truck right). Compare to *pan*.

Tuner/Timer The component that allows you to record by "tuning in to" broadcast television programs with a *VCR*. The "tuner" is a channel selector that allows you to tune in to the channel you want to record. The timer is a clock you can set to start and stop the VCR automatically.

TV Monitor See *monitor*.

TV Receiver A television set that has a tuner, so it can receive UHF and VHF television channels. Compare to *monitor*. See *RF* (*radio frequency*) and *tuner/timer*.

Unidirectional A microphone *pickup pattern*. Unidirectional mikes are sensitive to sound coming from only one direction—usually directly in front of the mike. Compare to *cardioid* and *omnidirectional*.

U-Type The *standard* three-quarter inch videocassette recorder format (the format that uses three-quarter-inch-wide tape. The "U" refers to the U-shaped pattern of the *tape path* in most three-quarter-inch VCRs.

VCR (videocassette recorder) The type of *VTR* that uses tape *cassettes* rather than separate open reels to hold and wind the tape. Some common types of VCR include *Beta*, *VHS*, and *U-Type*. Compare to *open-reel VTR*.

Vertical Sync The pulse in the *synchronization signal* that controls the rate at which the scanning beam switches lines as it scans the *raster* or *target area*. Compare to *horizontal sync*.

VHS (Video Home System) The half-inch videocassette *format* developed by Japan Victor Corporation (JVC). VHS cassettes are slightly larger and generally hold more tape than *Beta* format cassettes.

Video **1.** The picture portion of a TV signal. 2. Any electronic equipment used for producing, recording, transmitting, and displaying sounds and images on a TV set.

Videocassette See *cassette*.

Videocassette Recorder See *VCR*.

Videodisc Player A relatively new home video device that plays TV programs and movies from a flat disc that looks somewhat like a record album.

Video Head Drum The round metal casing that houses the spinning *video heads* in *helical-scan VTR s*.

Video Heads The small *electromagnets* inside a video recording deck that record and play back picture (video) information on a videotape. See also *helical scan*.

Videotape A special *magnetic tape* designed for recording and playing back video signals. Videotapes come in a variety of lengths, widths, and *formats*. See *cassette* and *ferrous oxide*.

Videotape Recorder See *VTR*.

Video Track The strip on the videotape where picture information is recorded. Compare to *audio track* and *sync track*.

Vidicon Tube A kind of *pickup tube* used in many home video cameras.

Viewfinder The device in or attached to a video camera that allows the camera operator to see the scene that is being recorded. See *electronic viewfinder*, *optical viewfinder*, and *through the lens (TTL) viewfinder*.

VTR (Videotape Recorder) Any machine that uses *magnetic tape* to record and play back television signals. The term VTR includes machines that house the tape in plastic *cassettes (VCRs)* as well as *open-reel* machines. See *deck*.

VTR Operator The *crew* member responsible for operating the VTR's *function controls* during a recording session.

White Balance A control on color video cameras that adjusts the camera's circuits for different lighting situations.

Wide-Angle Lens A camera lens with a short *focal length*. Wide-angle lenses take in a broad area of a scene. Contrast with *telephoto lens*. See *zoom lens*.

Wide-Angle Setting The *zoom* control setting that gives you the broadest (widest) view of a scene. At the wide-angle setting, the zoom is set at its shortest *focal length*. Compare to *maximum telephoto setting*.

Wide Shot A camera shot taken with the *zoom* control at or close to the *wide-angle setting*. See *establishing shot* and *long shot*.

Writing Speed See *tape-to-head speed*.

Zoom Making the subject seem closer (zoom in) or farther away (zoom out) by turning the zoom lens control.

Zoom Lens A camera lens with an adjustable *focal length*. A zoom lens is a single lens that can be adjusted to work as a *telephoto lens*, a *wide-angle lens*, and a normal lens.

index

About the Authors

John LeBaron is director of Massachusetts Educational Television, responsible for MET's project to plan for nonbroadcast means to distribute instructional television to Massachusetts schools. Mr. LeBaron received his M.Ed. and Ed.D. in Educational Media at the University of Massachusetts in Amherst. He has taught at the elementary, secondary, and college levels in Canada, Europe, and the United States. He currently resides with his family in Acton, Massachusetts.

Philip Miller is a telecommunications specialist at Massachusetts Educational Television, where he is involved in MET's research investigation of nonbroadcast alternatives for distributing educational programming. Mr. Miller received his master's degree in Television and Human Development from the Harvard Graduate School of Education. He has taught English and media studies at the high school level and communications courses at Tufts University and the University of British Columbia. He lives in Lincoln, Massachusetts.